This book is to be returned on or before

Space Science

HOW THE UNIVERSE WORKS

How to use this book

Welcome to *Space Science*. All the books in this set are organized to help you through the multitude of pictures and facts that make this subject so interesting. There is also a master glossary for the set on pages 58–64 and an index on pages 65–72.

The text is organized into chapters.

Capitals show key glossary terms. They are defined in the quick reference glossary.

Links to related information in other titles in the *Space Science* set.

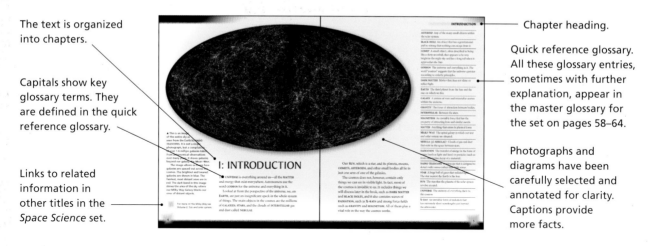

Chapter heading.

Quick reference glossary. All these glossary entries, sometimes with further explanation, appear in the master glossary for the set on pages 58–64.

Photographs and diagrams have been carefully selected and annotated for clarity. Captions provide more facts.

 Atlantic Europe Publishing

First published in 2004 by
Atlantic Europe Publishing Company Ltd.

Copyright © 2004
Atlantic Europe Publishing Company Ltd.
First reprint 2004

Author
Brian Knapp, BSc, PhD

Art Director
Duncan McCrae, BSc

Senior Designer
Adele Humphries, BA, PGCE

Editors
Mary Sanders, BSc, and Gillian Gatehouse

Illustrations on behalf of Earthscape Editions
David Woodroffe and David Hardy

Design and production
EARTHSCAPE EDITIONS

Print
WKT Company Limited, China

This product is manufactured from sustainable managed forests.
For every tree cut down, at least one more is planted.

Space science – Volume 1: How the universe works
A CIP record for this book is available from the British Library

ISBN 1 86214 363 3

Picture credits
All photographs and diagrams NASA except the following:
(c=center t=top b=bottom l=left r=right)
Earthscape Editions 6t, 6b, 8, 9, 20–21, 55; *ESO* 35, 45; *Jason Ware* 40–41; *ESA Artist* front cover; *NASA Artist* 36, 38–39.

The front cover shows an artist's impression of a black hole in a strong magnetic field; the back cover, different stages of the life cycle of stars in Nebula NGC 3603.

NASA, the U.S. National Aeronautics and Space Administration, was founded in 1958 for aeronautical and space exploration. It operates several installations around the country and has its headquarters in Washington, D.C.

CONTENTS

▲ The M27 Dumbbell Nebula.

▲ This is an image of the entire sky as seen from the Earth by **RADIO TELESCOPES**. It is not a single photograph, but a compilation of over 1.6 million galaxies taken from astronomical observations over many years. It shows galaxies beyond our galaxy, the **MILKY WAY**.

The image allows us to see how galaxies are spaced out across the cosmos. The brightest and nearest galaxies are shown in blue. The faintest, most distant ones are in red. The dark band in this image shows the area of the sky where our Milky Way Galaxy blocks our view of distant objects.

For more on the Milky Way see Volume 2: *Sun and solar system*.

1: INTRODUCTION

The **UNIVERSE** is everything around us—all the **MATTER** and energy that exist everywhere. Astronomers use the word **COSMOS** for the universe and everything in it.

Looked at from the perspective of the universe, we, on **EARTH**, are just an insignificant speck in the whole system of things. The main objects in the cosmos are the millions of **GALAXIES**, **STARS**, and the clouds of **INTERSTELLAR** gas and dust called **NEBULAE**.

Our **SUN**, which is a star, and its planets, moons, **COMETS**, **ASTEROIDS**, and other small bodies all lie in just one arm of one of the galaxies.

The cosmos does not, however, contain only things we can see in visible light. In fact, most of the cosmos is invisible to us. It includes things we will discuss later in the book, such as **DARK MATTER** and **BLACK HOLES**, and it also contains waves of **RADIATION**, such as **X-RAYS** and strong force fields such as **GRAVITY** and **MAGNETISM**. All of them play a vital role in the way the cosmos works.

ASTEROID Any of the many small objects within the solar system.

BLACK HOLE An object that has a gravitational pull so strong that nothing can escape from it.

COMET A small object, often described as being like a dirty snowball, that appears to be very bright in the night sky and has a long tail when it approaches the Sun.

COSMOS The universe and everything in it. The word "cosmos" suggests that the universe operates according to orderly principles.

DARK MATTER Matter that does not shine or reflect light.

EARTH The third planet from the Sun and the one on which we live.

GALAXY A system of stars and interstellar matter within the universe.

GRAVITY The force of attraction between bodies.

INTERSTELLAR Between the stars.

MAGNETISM An invisible force that has the property of attracting iron and similar metals.

MATTER Anything that exists in physical form.

MILKY WAY The spiral galaxy in which our star and solar system are situated.

NEBULA (pl. **NEBULAE**) Clouds of gas and dust that exist in the space between stars.

RADIATION The transfer of energy in the form of waves (such as light and heat) or particles (such as from radioactive decay of a material).

RADIO TELESCOPE A telescope that is designed to detect radio waves rather than light waves.

STAR A large ball of gases that radiates light. The star nearest the Earth is the Sun.

SUN The star that the planets of the solar system revolve around.

UNIVERSE The entirety of everything there is; the cosmos.

X-RAY An invisible form of radiation that has extremely short wavelengths just beyond the ultraviolet.

Origins of the universe

We now think we have a reasonable idea of how the universe works, when it was formed, and what will happen to it in the future. This will be set out in more detail later in the book. But we are not the first people (and probably not the last) to feel that they know the answer to the fundamental question of where we are.

Since earliest times people have wondered about the universe they live in. Many people have believed through the millennia, and still do today, that the universe has a supernatural cause.

The ancient Greeks were among the first to think that the universe had developed naturally. Their reasoning for this was that wherever they looked, they saw a universe that seemed to obey the laws of mathematics. It was, in effect, an ordered universe. This, they thought, could easily have come about by natural means.

They also developed a startlingly perceptive theory that the universe was due to the interplay of **ATOMS** coming together or breaking apart and creating endless worlds in various stages of development and decay.

By the 4th century B.C. one Greek thinker, Heracleides, was already teaching that the Earth rotated freely in space and that Mercury and Venus revolved round the Sun.

The ancient Greeks also knew that the Sun is much larger than the Moon even though both appear to be the same size from Earth. Aristarchus suggested that the Earth revolves around the Sun, and not the Sun around the Earth.

▲▼ Our knowledge of the universe has been greatly improved by newer telescopes that can see farther and more clearly.

Copernicus to Newton

After the Greeks and Romans mathematical theories were lost for many centuries, and the Earth was once again thought of as flat and as the center of the universe.

It was not until the 16th century that observations by scientists such as Copernicus, Tycho Brahe, Kepler, and Galileo opened the minds of scientists once more to the fact that the Sun was the center of the **SOLAR SYSTEM** and that it was but one part of the universe. This was given support by the calculations of Sir Isaac Newton, who developed the important **LAWS OF MOTION**.

Observing stars

During the 1700s, ever more powerful telescopes were built. That allowed people such as the British astronomer William Herschel and his son John to measure enormous numbers of faint stars and to try to find how they are placed in the universe. Herschel was the first to realize that stars are enormous distances apart.

Later, photographic evidence would show that stars are not truly fixed in position, but move very slowly across the sky. They only appear unmoving because of their enormous distances from us.

The sky at night

One of the problems in understanding how the universe works is the need to get a clear view of where everything is. Although since earliest times people have been able to look at the sky and see stars and some planets, they have always seen them from the same position. That can produce a distorted view of the real shape and position of objects in space.

For example, we see a band of concentrated stars sweeping over the sky, and we call them the **MILKY WAY**. But they are not a band. Instead, they are the edge of a flattened disk containing countless stars. From Earth we cannot see the true shape of the **GALAXY** of stars in the Milky Way.

Similarly, when we look at our Sun, it is not at all obvious to us that it is just like the stars we see at night. That is because it is so close that we see it as a yellow-white colored disk. The nearest star is so far away that we cannot see it as a disk at all, but only as a point of light. This is the same effect as we get when we see car headlights far off at night. They seem to be just two points of light; but as the car gets closer, we can see the twin headlights as two yellow-white disks.

Many features of the universe are like this. For example, many of the largest structures in the observable universe are so far away that they can only be seen as points or blurs of light. And if they do not shine with their own internal light (which includes X-rays and **RADIO WAVES**), we do not see (detect) them at all.

ATOM The smallest particle of an element.

GALAXY A system of stars and interstellar matter within the universe.

HUBBLE SPACE TELESCOPE An orbiting telescope (and so a satellite) that was placed above the Earth's atmosphere so that it could take images that were far clearer than anything that could be obtained from the surface of the Earth.

LAWS OF MOTION Formulated by Sir Isaac Newton, they describe the forces that act on a moving object.

MILKY WAY The spiral galaxy in which our star and solar system are situated.

RADIO WAVES A form of electromagnetic radiation, like light and heat. Radio waves have a longer wavelength than light waves.

SENSOR A device used to detect something.

SOLAR SYSTEM The Sun and the bodies orbiting around it.

▼ The **HUBBLE SPACE TELESCOPE**, which took most of the images in this book, does not "see" space. Its **SENSORS** detect a range of wavelengths of radiation other than light. The colors shown in its images are not always what we would see if we were to go into space. Instead, colors are used to represent temperature, for example, or they can display wavelengths of radiation that we would never see at all.

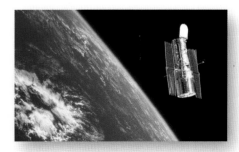

Looking at the night sky

We can all see patterns in the stars that dot the night sky. Examples include the Great Bear (Ursa Major), Orion, and the Southern Cross (Crux).

Some people have believed that the patterns represent gods or animals and other features. But in any case, the star groupings are one way of classifying objects in the sky. Major star groupings are called CONSTELLATIONS.

Stars are very distant objects, and their position only changes very slightly over hundreds of years. They therefore appear to be fixed in the sky. Travelers have long used constellations for navigating both continents and oceans.

▼ Constellations in the northern sky.

N

Ursa Minor

Draco

Auriga

Lynx

Ursa Major

Boötes

Corona Borealis

Canes Venatici

Leo Minor

Gemini

Hercules

Cancer

Coma Berenices

Leo

Canis Minor

Orion

Serpens Caput

Sextans

Canis Minor

E

Ophiuchus

Virgo

Crater

Hydra

Monoceros

W

Corvus

Libra

Serpens Cauda

Hydra

Pyxis

Canis Major

Antilia

Scorpius

Centaurus

Vela

Puppis

Columba

Lupus

Crux

Norma

Carina

Circinus

Musca

Ara

Triangulum Australe

Pictor

Pavo

Chamaeleon

Volans

Apus

Octans

Mensa

S

For centuries constellations have also been used as **CELESTIAL** timekeepers, their position in the sky telling, for example, when to plant crops.

Some objects in the sky do not, however, keep the same position, but move from day to day. The Greek word for wanderers is *planetes*, from which we get the modern word "planet." Mercury, Venus, Mars, Jupiter, and Saturn have all been known for thousands of years. Nowadays sky wanderers include some of the many **SATELLITES** we have put into **ORBIT** around the Earth.

CELESTIAL Relating to the sky above, the "heavens."

CONSTELLATION One of many commonly recognized patterns of stars in the sky.

ORBIT The path followed by one object as it tracks around another.

SATELLITE A man-made object that orbits the Earth.

▼ Constellations in the southern sky.

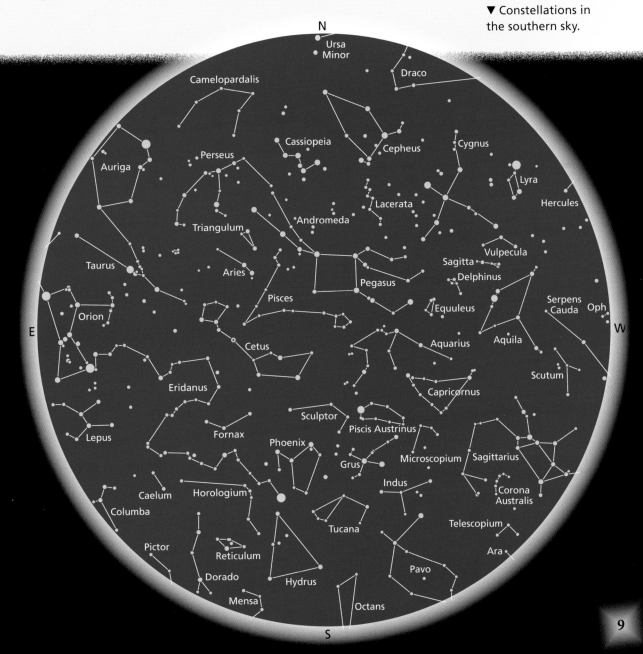

**Eta Carinae, the
keyhole nebula
(catalogue number NGC 3372)**

▲ Eta Carinae is 10,000 light-years away, and so we cannot see it in fine detail. We can only see features of about the size of our solar system, 16 billion km across.

Because it is unstable, it is prone to sudden, violent outbursts. One eruption occurred about 150 years ago, and then Eta Carinae became one of the brightest stars in the southern sky. The explosion produced two lobes and a large, thin equatorial disk.

Eta Carinae has a mass of about 100 to 150 Suns and is one of the most massive and most luminous stars known. It is about four million times brighter than our Sun.

Measuring the heavens

To help get a perspective on the objects in the universe, we need a CELESTIAL measure. But our usual measurement system of kilometers will not do for such large distances. We need something more appropriate. Three units are used, in part depending on the distances involved:

Astronomical unit

The basic measurement for the solar system is the ASTRONOMICAL UNIT (AU). It is the average distance from the Earth to the Sun (149,597,870 km).

Light-year

The distance traveled by light through space in one Earth year is called a LIGHT-YEAR. A light-year equals about 9,460,000,000,000 km, or 63,240 astronomical units.

Light travels through space at 300,000,000 meters per second. For measuring very long distances, therefore, the time it takes for light to cover a distance is an ideal measure.

Light travels around the Earth seven times a second, so it is not a useful measurement on Earth; it takes 8 minutes for light to reach us from the Sun, so that is not a very useful measurement even for distances within the solar system. By contrast, the time it takes for light to travel from the nearest star outside our solar system, Proxima Centauri, is 4.3 years. Most stars are vastly farther away. So, while it makes sense to use AUs for the solar system, it is more manageable to use light-years for **INTERSTELLAR** distances.

The time it would take for light to travel from one side of a **GALAXY** to another is, on average, 100,000 years, that is, a galaxy is 100,000 light-years across.

Our galaxy has a hundred billion stars in it. The galaxies nearest to our Milky Way are the small ones called the Large and Small Magellanic Clouds (see page 51). They are 200,000 light-years away.

The nearest big galaxy is the Andromeda Galaxy (also called M31). It is two million light-years away. That means light that left the Andromeda Galaxy before the start of the Earth's Ice Age is only now just reaching us.

Parsec

The **PARSEC** is used by professional astronomers to measure the largest distances in the universe, such as the distances to galaxies.

It is based on calculation of the **RADIUS** of the Earth's **ORBIT** round the Sun. A parsec is the distance from which an observer in space would see the radius of the orbit as making 1 second of arc ($^1/_{3600}$th of a degree). This distance is about 3.26 light-years. A kiloparsec is a thousand parsecs. The Sun, for example, is 8.5 kiloparsecs from the center of the Milky Way.

A megaparsec is a million parsecs. The distance to the Andromeda Galaxy is 0.7 megaparsec. Some galaxies may be up to 3,000 megaparsecs away.

**Lagoon Nebula
(catalogue numbers NGC 6523
and M8)**
▲ This area of hydrogen gas in the constellation Sagittarius is 4,000 light-years away and about 33 light-years across.

ASTRONOMICAL UNIT (**AU**) The average distance from the Earth to the Sun (149,597,870 km).

CELESTIAL Relating to the sky above, the "heavens."

GALAXY A system of stars and interstellar matter within the universe.

INTERSTELLAR Between the stars.

LIGHT-YEAR The distance traveled by light through space in one Earth year, or 63,240 astronomical units.

ORBIT The path followed by one object as it tracks around another.

PARSEC A unit used for measuring the largest distances in the universe.

RADIUS (pl. **RADII**) The distance from the center to the outside of a circle or sphere.

▶ M15 Globular cluster in the constellation Pegasus.

Cataloguing objects in the sky

It is easy to use names when you are talking about a small number of things, such as planets in the solar system. But when you extend your study to every object in the night sky, the task of describing them is much more complex, and a special reference system is needed. Two systems have been evolved: an earlier one, called the Messier system, and a more recent one, called the New General Catalogue (NGC).

The Messier Catalogue

At the end of the 18th century Charles Messier made a list of 109 **STAR CLUSTERS**, **GALAXIES**, and **NEBULAE**. Each object was given an **M NUMBER** based on the order in which he catalogued them. The Crab Nebula, our nearest major galaxy, is M1 (see pages 13 and 32), the Andromeda Galaxy is M31 (see pages 40–41), and so on. Note that only the most visible objects have M numbers because the **RESOLVING POWER** of telescopes at the end of the 18th century was not as good as it is today.

New General Catalogue of nebulae and star clusters (NGC)

This cataloguing approach was developed by the Danish astronomer Johan Ludvig Emil Dreyer in 1888. He based it on information on stellar objects gathered by the Herschel family in Britain.

The NGC is an alternative to the Messier system for stellar objects. An object can be classified in both systems. Like Messier, it does not cover all the objects that can be seen with modern equipment.

GALAXY A system of stars and interstellar matter within the universe.

M NUMBERS In 1781 Charles Messier began a catalogue of the objects he could see in the night sky. He gave each of them a unique number. The first entry was called M1. There is no significance to the number in terms of brightness, size, closeness, or otherwise.

NEBULA (pl. **NEBULAE**) Clouds of gas and dust that exist in the space between stars.

RESOLVING POWER The ability of an optical telescope to form an image of a distant object.

STAR CLUSTER A group of gravitationally connected stars.

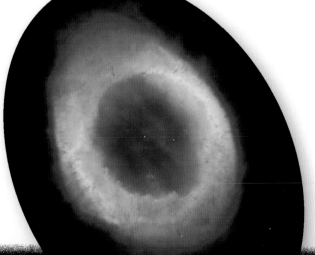

▶ M57 Ring Nebula in the constellation Lyra.

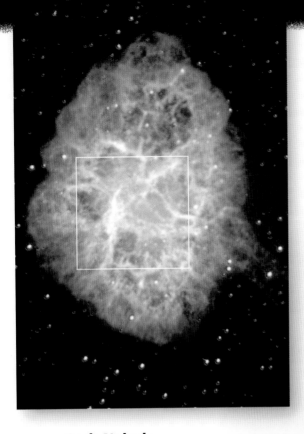

These pictures show a selection of objects with their Messier and their NGC numbers. You will find others in this book.

Dumbbell Nebula
▼ Planetary Nebula M27 (NGC 6853) in the constellation Vulpecula.

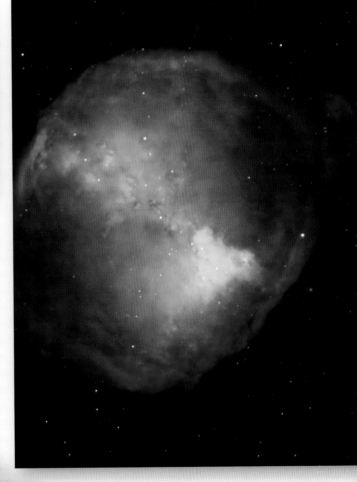

Crab Nebula
▲ Supernova Remnant M1 (NGC 1952) in the constellation Taurus (see page 32).

Orion Nebula
◄ Nebula M42 (NGC 1976) in the constellation Orion.

2: STARS

Although **STARS** are just tiny specks within the vastness of the universe, everything around us originates from them.

To understand this, we have to know that everything in the universe is made from the basic building blocks that we call **ATOMS**. Atoms combine to form **MOLECULES** in **ELEMENTS**. Hydrogen is an element, as are helium, calcium, sodium, and silicon. Hydrogen is the most common element in the universe. It is the basic element from which every other element is formed.

Every element has an **ATOMIC WEIGHT** based on the amount of **MATTER** in the core (nucleus) of each atom of the element. Hydrogen has the least matter in its atom and so is a very light element. Helium is the next lightest, while other elements are much heavier. As a result, we can think of light elements and heavy elements.

Star factories

Some people think of the dust and gas of the universe as a kind of raw material, with the stars as factories that process this material and send out new products—in this case new elements.

It takes stars a long time to complete this feat. Stars that are young create elements heavier than hydrogen but then keep them. Then, later on in the life of the star, when the hydrogen is used up, the star explodes and throws out all of the heavy elements as gas clouds that float around among the stars. This is called interstellar (between the stars) matter. Later it will be used to make new stars.

▶ In **NEBULA** NGC 3603 different stages of the life cycle of stars can be seen in one single view.

Here you can see Bok globules (the earliest stage of star formation) and giant gaseous pillars (which tell of **RADIATION** interacting with hydrogen to begin the formation of stars). In the center of the view there is a starburst cluster of young stars. They emit radiation in the form of **STELLAR WINDS** strong enough to blow a large hole around the cluster. The blue giant (Sher 25) (upper right) with its ring marks the end of the life cycle.

This is a true-color picture.

ATOM The smallest particle of an element.

ATOMIC WEIGHT The ratio of the average mass of a chemical element's atoms to carbon-12.

BLUE GIANT A young, extremely bright and hot star of very large mass that has used up all its hydrogen and is no longer in the main sequence. When a blue giant ages, it becomes a red giant.

ELEMENT A substance that cannot be decomposed into simpler substances by chemical means.

MATTER Anything that exists in physical form.

MOLECULE A group of two or more atoms held together by chemical bonds.

NEBULA (pl. **NEBULAE**) Clouds of gas and dust that exist in the space between stars.

RADIATION The transfer of energy in the form of waves (such as light and heat) or particles (such as from radioactive decay of a material).

STAR A large ball of gases that radiates light. The star nearest the Earth is the Sun.

STELLAR WIND The flow of tiny charged particles (called plasma) outward from a star.

◄▼ This **HUBBLE SPACE TELESCOPE** picture shows part of the Eagle Nebula (M16; NGC 6611) 7,000 **LIGHT-YEARS** from Earth in the **CONSTELLATION** Serpens. When we look at a picture like this, we see the universe as it was 7,000 years ago, not as it is today.

It shows newborn stars emerging from dense, compact pockets of interstellar gas called evaporating gaseous globules (EGGs).

The EGGs are at the tip of fingerlike features made of cold hydrogen gas and dust that stick out from the wall of a vast cloud of the same material. Each finger is several light-years long.

The interstellar gas is dense enough to collapse under its own weight, forming young stars that continue to grow as they suck in more and more mass from their surroundings.

The newborn stars send out a flood of ultraviolet light, causing much of the surrounding hydrogen to **EVAPORATE** (the process is called photoevaporation). The photoevaporation disperses the cloud made by the EGG and leaves the stars revealed as isolated objects in space.

Red shows light emitted by sulfur atoms. Green shows light emitted by hydrogen atoms. Blue shows light emitted by oxygen atoms.

The birthplace of new stars: interstellar clouds

Stars were created from material scattered through **SPACE**. At the end of their lives they will throw much of this material back into space. In this way the same material is used over and over again. It is called interstellar gas and dust. When it is concentrated enough to be seen, it is called an interstellar cloud. Dust makes up about 1% of these clouds and so contains much of the heavy elements in the universe.

The most common clouds are made from hydrogen atoms or molecules. Clouds made from **IONIZED** gas (gas in which the particles are charged) are less common but much more spectacular, containing **FLUORESCENT** masses of gas. The hydrogen atoms break apart and become ionized when they are bombarded by **ULTRAVIOLET** light (**PHOTONS**) from massive blue giant stars being formed within the cloud. Examples include the Orion Nebula.

CONSTELLATION One of many commonly recognized patterns of stars in the sky.

EVAPORATE The change in state from liquid to a gas.

FALSE COLOR The colors used to make the appearance of some property more obvious.

FLUORESCENT Emitting the visible light produced by a substance when it is struck by invisible waves, such as ultraviolet waves.

HUBBLE SPACE TELESCOPE An orbiting telescope (and so a satellite) that was placed above the Earth's atmosphere so that it could take images that were far clearer than anything that could be obtained from the surface of the Earth.

INFRARED Radiation with a wavelength that is longer than red light.

IONIZED Matter that has been converted into small charged particles called ions.

LIGHT-YEAR The distance traveled by light through space in one Earth year, or 63,240 astronomical units.

PHOTON A particle (quantum) of electromagnetic radiation.

SPACE Everything beyond the Earth's atmosphere.

ULTRAVIOLET A form of radiation that is just beyond the violet end of the visible spectrum and so is called "ultra" (more than) violet. At the other end of the visible spectrum is "infra" (less than) red.

◄ This star is 25,000 light-years away in the direction of the Sagittarius constellation. It is one of the brightest stars in our galaxy, though it is normally hidden from us by cosmic dust. The picture was taken using only **INFRARED** light, which can penetrate the dust cloud. The star has a radiance of 10 million Suns.

In this picture you can also see a massive star explosion. The explosion has sent enough material into space to create two shells of gas (red color) with a mass equal to several of our Suns. The largest shell is 4 light-years across.

This star is shedding material at a tremendous rate and may once have had a mass of 200 of our Suns.

This is a **FALSE-COLOR** image.

Galaxy NGC 6822

▲ This is a glowing gas cloud, called Hubble-V, with a diameter of about 200 light-years. There is a dense knot of dozens of ultrahot stars nestled in the nebula, each glowing 100,000 times brighter than our Sun. They are youthful four-million-year-old stars, forming part of a small, irregular host galaxy called NGC 6822, one of the closest neighbors to the **MILKY WAY**. The galaxy is 1.6 million light-years away in the constellation Sagittarius.

Most stars are made of clouds of molecular material. A typical molecular cloud might contain enough matter to make a million stars the size of our Sun. Even though it is hundreds of light-years across in size, the cloud has enough **GRAVITY** to cause some matter to begin to group together. Most molecular clouds contain clumps of material, and it is in these cloud cores that stars develop.

In a cloud a clump of material starts to spin slowly. Eventually the clumps get big enough to develop sufficient gravity for them to begin to collapse in on themselves. Almost invisible matter now suddenly becomes a concentrated and opaque ball of gas called a **PROTOSTAR**.

Once star formation is under way, and the new star is still perhaps not much bigger than one of our gas giant planets, the gravity of the protostar causes more material to move toward it. More pressure raises the temperature of the core, and that makes light begin to be radiated from it. At first the star is not very hot and is only visible as an infrared object. It cannot be seen through an **OPTICAL** telescope.

The protostar becomes more and more active, sending out jets of material as a **STELLAR WIND**. In time its stellar wind blocks any material that might be moving toward it, and because of this the star eventually stops growing.

As protostars heat up, they become visible for the first time and are called T-Tauri stars (for a picture see page 48). Unlike most newly born objects, these new stars are actually bigger than they will be as they mature, because gravity has not yet had time to pull them down to their final size. When that does happen, hydrogen **FUSION** starts up in the core, and the protostar becomes a **MAIN-SEQUENCE** star (see page 24).

Binary stars

About half of all stars are found in pairs in orbit around their common **CENTER OF GRAVITY**. Sometimes one star will consume the material of the other and cause an intense increase in brightness. This will be discussed under nova and supernova (see pages 30–31).

Classifying a star

Grouping, or classifying, stars helps identify patterns both in the stars and in the galaxies that contain them.

There are many ways in which stars can be grouped. For example, astronomers label young stars (those less than 100 million years old and which still contain heavy elements such as iron, nickel, and carbon) as Population I stars. The Sun is a Population I star. All of these kinds of stars occur in the arms of spiral galaxies. By contrast, stars that are older contain little by way of heavy elements because, in the earliest years of the universe, such elements had not yet been created. They are called Population II stars. They occur in the globular clusters at the heart of spiral galaxies and throughout elliptical galaxies.

Each new generation of stars contains a higher proportion of heavy elements than its predecessor. That is why, over time, the universe gradually contains less hydrogen and more of every other element. The gas clouds emitted by Population I stars will eventually be used again to form a new generation of stars.

Another way of classifying stars is through their size and color.

The size of a star is often related both to how bright it is and to its color. Small stars like the Sun are a few thousand degrees Celsius at their surfaces. They are yellow. Large stars are several tens of thousands of degrees at the surface because they are burning more fiercely. As a result, they shine with a blue-white light.

Astronomers classify the hottest stars as type O and the coldest as M. A complete sequence of categories is O B A F G K M. The Sun ranks as a G star.

Stars can also be thought of as having a life cycle, as we shall see from page 20. As a result, it is useful to describe where stars are in their life. The numbers I (for the largest and youngest stars) to V (for the old **DWARF STARS**) are used for this.

Star clusters

Stars are not isolated in their galaxies. It is common to find groups of stars bound together by their own gravity. They are called **STAR CLUSTERS**. There are two types of star clusters: open clusters and globular clusters.

Open clusters have a hundred to a thousand stars. There does not appear to be any regular organization to an open cluster.

Globular clusters are simply vast. They can contain hundreds of thousands, or millions, of stars, all concentrated together.

In both types of cluster the stars were formed at the same time from the same cloud of gas and dust. As they developed, some grew faster than others, gathering more matter to themselves and depriving others of the chance to grow.

Because big stars evolve so much more rapidly than small stars (see page 23), some stars in a cluster have gone farther through their life cycles than others.

CENTER OF GRAVITY The point at which all of the mass of an object can be balanced.

DWARF STAR A star that shines with a brightness that is average or below.

FUSION The joining of atomic nuclei to form heavier nuclei.

GRAVITY The force of attraction between bodies.

MAIN SEQUENCE The 90% of stars in the universe that represent the mature phase of stars with small or medium mass.

MILKY WAY The spiral galaxy in which our star and solar system are situated.

OPTICAL Relating to the use of light.

PROTOSTAR A cloud of gas and dust that begins to swirl around; the resulting gravity gives birth to a star.

STAR CLUSTER A group of gravitationally connected stars.

STELLAR WIND The flow of tiny charged particles (called plasma) outward from a star.

Globular clusters are the oldest star clusters, averaging about 15 billion years. So these clusters are some of the oldest objects in the universe.

The life and death of stars

Stars do not last for the entire existence of the universe. Since the universe began, many generations of stars have been formed and eventually have faded away. As a result, it is helpful to think of stars as having a life cycle. The life of a star can be divided into birth (rapid change), adulthood (period of stability), and decay or death.

The birth of a star

The universe contains a vast amount of gas and dust in the form of interstellar matter, which can be made into stars. It is simply a matter of enough of it coming together to begin a CHAIN REACTION.

It is thought that waves of RADIATION passing through the universe stir up the interstellar gas and dust clouds, making them swirl around and so become more concentrated.

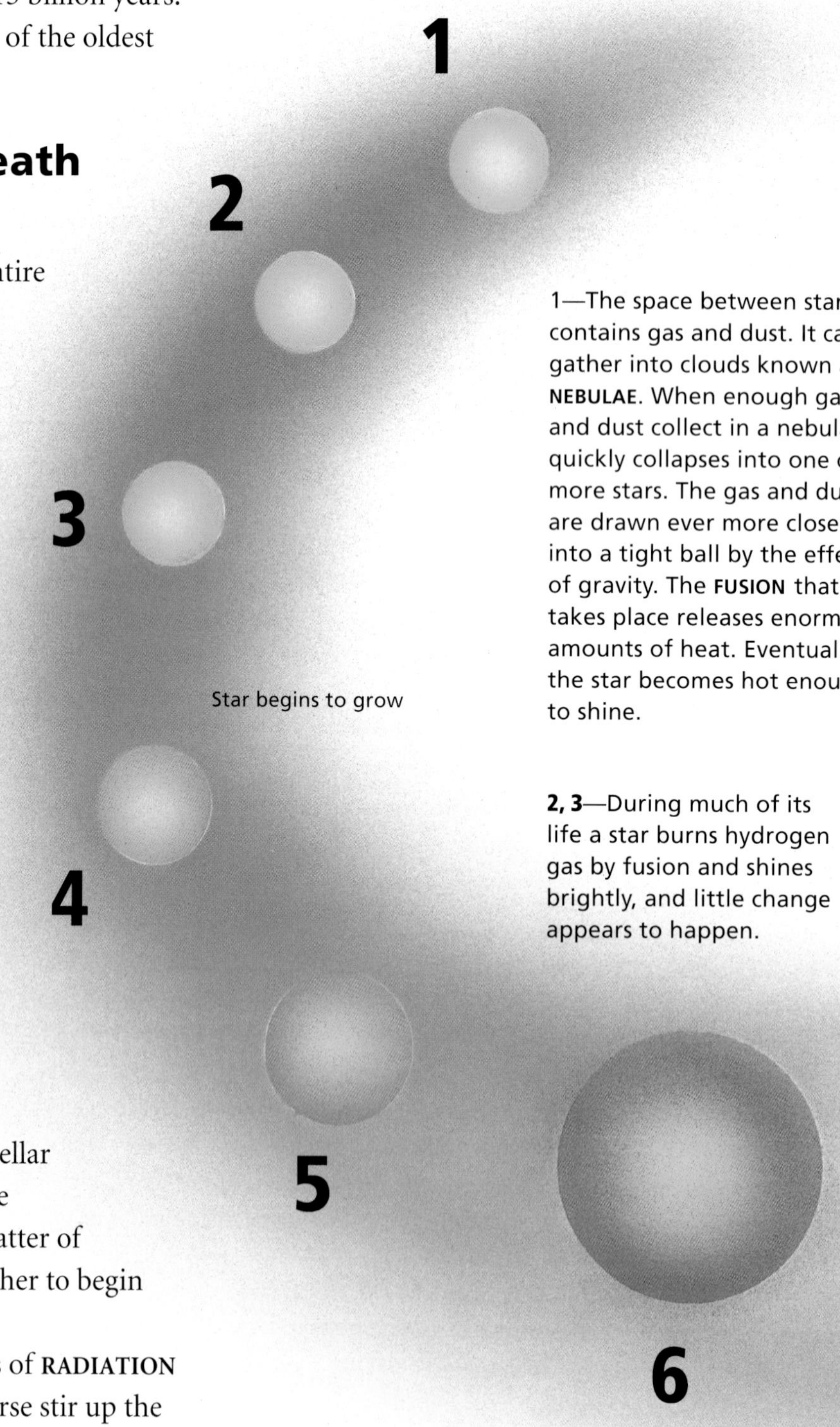

1

2

3

Star begins to grow

4

5

6

1—The space between stars contains gas and dust. It can gather into clouds known as NEBULAE. When enough gas and dust collect in a nebula, it quickly collapses into one or more stars. The gas and dust are drawn ever more closely into a tight ball by the effects of gravity. The FUSION that takes place releases enormous amounts of heat. Eventually, the star becomes hot enough to shine.

2, 3—During much of its life a star burns hydrogen gas by fusion and shines brightly, and little change appears to happen.

4, 5, 6, 7, 8—Eventually, the hydrogen fuel is used up, and only helium is left to burn. Since helium burns at much higher temperatures than hydrogen, the star gets brighter. At the same time, the outer part of the star begins to expand again, forming a **RED GIANT** star (page 26).

Eventually, the star literally blows apart and produces the spectacular "fireworks" in space called a **SUPERNOVA** (pages 31–33).

A supernova is a red giant that explodes. It suddenly increases in brightness by a factor of many billions, but even within a few weeks it begins to fade. The Crab Nebula (some 7,000 light-years away) consists of material ejected by the supernova of 1054 (for pictures see pages 13 and 32).

A supernova may radiate more energy in a few days than the Sun does in 100 million years. The stellar remnant left behind after the explosion is a star only a few kilometers in diameter but with an enormously high density.

BLACK DWARF A degenerate star that has cooled so that it is now not visible.

CHAIN REACTION A sequence of related events with one event triggering the next.

FUSION The joining of atomic nuclei to form heavier nuclei.

NEBULA (pl. **NEBULAE**) Clouds of gas and dust that exist in the space between stars.

NEUTRON STAR A very dense star that consists only of tightly packed neutrons. It is the result of the collapse of a massive star.

PULSAR A neutron star that is spinning around, releasing electromagnetic radiation, including radio waves.

RADIATION The transfer of energy in the form of waves (such as light and heat) or particles (such as from radioactive decay of a material).

RADIO WAVES A form of electromagnetic radiation, like light and heat. Radio waves have a longer wavelength than light waves.

RED GIANT A cool, large, bright star at least 25 times the diameter of our Sun.

SUPERNOVA A violently exploding star that becomes millions or even billions of times brighter than when it was younger and stable.

WHITE DWARF Any star originally of low mass that has reached the end of its life.

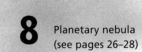

11 Black dwarf

10

9 White dwarf (neutron star, pulsar)

8 Planetary nebula (see pages 26–28)

9, 10, 11—The remnant star then contracts as a **NEUTRON STAR** or **WHITE DWARF**, spinning quickly and sending out pulses of **RADIO WAVES** like a galactic beacon. That is why a neutron star is also sometimes called a **PULSAR** (page 34). Eventually, the neutron star dies, its heat is lost, and it ceases to shine. Although it still exists in the galaxy, it can no longer be seen, and it is now called a **BLACK DWARF** (page 29).

7 Red giant

◀ These are hot blue stars deep inside an elliptical (oval) galaxy seen in ultraviolet light. This picture shows 8,000 blue stars in galaxy M32, some 2.5 million light-years away in the constellation Andromeda.

The ultraviolet light comes from extremely hot, helium-burning stars at a late stage in their lives. Unlike the Sun, which burns hydrogen into helium, these old stars exhausted their central hydrogen long ago and now burn helium into heavier elements.

▲ This is the "polar-ring" galaxy NGC 4650A, located about 130 million light-years away. The bright bluish clumps, which are especially prominent in the outer parts of the ring, are regions containing shining young stars.

Stars are born when enough interstellar matter comes together for a large **GRAVITATIONAL FIELD** to develop. As **GRAVITY** pulls the matter (which is mostly hydrogen) together, it heats up the gas so much that nuclear reactions take place. That generates more heat, and the star begins to shine. This is the birth of a star.

As hydrogen burns, it is converted to helium. During this conversion four hydrogen nuclei change into one helium nucleus. But not all of the nuclear energy that bound the hydrogen atoms is required to form a helium nucleus, and so some is released as radiation. Among the forms of radiation released are heat and light, which make a star shine brightly.

Mature stars

Not all stars have the same life. Stars that are smaller than the Sun burn hydrogen much more slowly than a star that is bigger than the Sun. As a result, small stars have a longer life than their large, hydrogen-eating cousins.

A star about half the size of the Sun cannot generate the massive burning that occurs in a big star. Because of this it burns its fuel more slowly and lasts for a long time, often billions of years, as our Sun demonstrates. On the other hand, a star 50 times the size of the Sun can process hydrogen much more quickly and spectacularly. As a result, although it contains more hydrogen than a small star, it uses it up faster and lasts for only three million years.

Small stars can last as long as galaxies. Big, bright blue stars, on the other hand, with lifetimes ten thousand times shorter than small stars, come and go very fast during the lifetime of a galaxy, being born among the clouds of gas and dust in the galaxies. You can see this happening in many pictures in this book.

GRAVITATIONAL FIELD The region surrounding a body in which that body's gravitational force can be felt.

GRAVITY The force of attraction between bodies.

Main-sequence stars

The longest part of the life of a star is when it is mature. It has formed enough **GRAVITY** for nuclear reactions to be well under way, but it still has enough hydrogen fuel left to keep it burning steadily. This long period of stability is referred to as the **MAIN SEQUENCE**. It is when the star shines steadily, and when it transforms hydrogen to helium in its core.

About nine out of ten stars are in the main sequence, and every one of them is a **DWARF STAR**. A star is only classified as a larger size as it swells toward the end of its life.

This means that most of the **MASS** of a galaxy is in main-sequence stars. In fact, most stars are small mass stars. But since they are not as bright as large stars, they are not as conspicuous. As a result, more starlight is produced by the larger stars than you would think by simply comparing star mass.

Small stars do become brighter as they age. The Sun is becoming brighter in this way. Even later, near the end of their lives, stars may become cooler and redder.

Old age for a star

A star is a mass of swirling, burning gas. Gas is easily compressed. The great mass of a star ensures that it has enormous gravity. Gravity therefore tends to collapse the star in on itself all the time. But for much of its life the star burns so fiercely that the gases are constantly expanding. A balance is set up between expansion due to burning and the squashing effect of gravity. This lasts all of the time the star is in the main sequence.

The star begins to die when there isn't enough fuel in the form of hydrogen to maintain the nuclear reactions needed to keep the gases burning and expanding.

Stars, however, do not all end the same way. Just as at birth, what happens at death depends on size.

For more on our Sun see Volume 2: *Sun and solar system*.

DWARF STAR A star that shines with a brightness that is average or below.

GRAVITY The force of attraction between bodies.

MAIN SEQUENCE The 90% of stars in the universe that represent the mature phase of stars with small or medium mass.

MASS The amount of matter in an object.

STELLAR WIND The flow of tiny charged particles (called plasma) outward from a star.

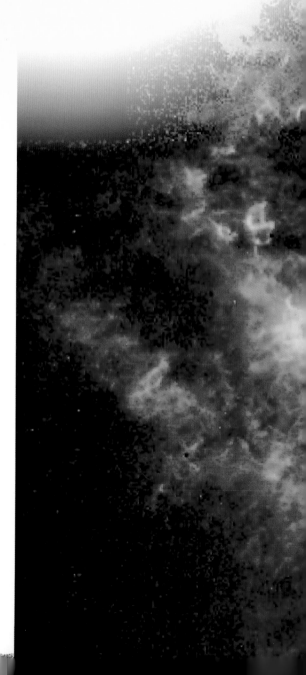

▼ This is a picture of the energetic hot star WR124 ejecting hot clumps of gas into space at speeds of over 160,000 kilometers an hour.

The massive star, called a Wolf-Rayet star, is a short-lived class of superhot star ejecting mass to make a furious **STELLAR WIND**.

The surrounding nebula is estimated to be no older than 10,000 years.

The star is 15,000 light-years away, located in the constellation Sagittarius.

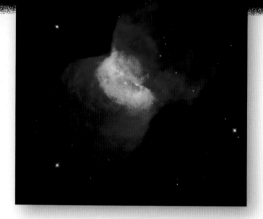

Butterfly Nebula

▲ This is the Butterfly Nebula NGC 2346. The nebula is about 2,000 light-years away from Earth in the direction of the constellation Monoceros.

It represents the spectacular "last gasp" of a binary star system at the nebula's center. At the center of the nebula lies a pair of stars that are so close together that they orbit around each other every 16 days. This is so close that they appear to be just one star. One star is the hot core of a star that has ejected most of its outer layers, producing the surrounding nebula.

Astronomers believe that this star, when it evolved and expanded to become a red giant, actually swallowed its companion star in an act of stellar cannibalism. The resulting interaction led to a spiraling together of the two stars, resulting in ejection of the outer layers of the red giant. Later, the hot star developed a fast stellar wind. This wind, blowing out into the surrounding disk, has inflated the large, wispy hourglass-shaped wings perpendicular to the disk. The total diameter of the nebula is about one-third of a light-year, or three trillion kilometers.

CONDENSATION To make something more concentrated or compact.

RED GIANT A cool, large, bright star at least 25 times the diameter of our Sun.

STELLAR WIND The flow of tiny charged particles (called plasma) outward from a star.

Red giant stars

Small stars do not just fade away as they die. As the fuel begins to run out in a small star, the gases cannot oppose gravity, and the core contracts. That makes it heat up, and for a while even more energy reaches the outside. As a result, the last phase of a star is not a slow event, but one that is often dramatic.

As the outer regions of the star become heated, they expand. At this stage the star turns from a main-sequence yellow star and grows into a **RED GIANT** star.

At the same time, the nuclear "ash" at the core, which is made of helium, may get so hot that the helium atoms fuse into carbon. Then, if the temperature is high enough, carbon and helium atoms can fuse to produce oxygen. So the death of a star also produces new elements.

Meanwhile, if the shell of the star has burned away all of its hydrogen to helium, the helium can then begin to fuse in this region too, and the whole process of creating heavier elements can also take place in the outer shell of the star.

The outer atmospheres of such a star may be cool enough to allow the **CONDENSATION** of some of the heavy elements into solid particles, creating grains of dust.

Planetary nebula

The force of gravity depends on a concentration of mass. With an enormously expanded shell of gases and dust, gravity becomes too weak to hold much of the dead star together, and much of it blows out of the star to form a violent **STELLAR WIND** (the same kind of thing as the solar wind, but vastly greater in amount).

▶ This is a picture of a dying star, NGC 6543, nicknamed the "Cat's Eye Nebula." You can see concentric gas shells, jets of high-speed gas, and unusual shock-induced knots of gas.

A fast stellar wind of gas blown off the central star created the elongated shell of dense, glowing gas. NGC 6543 is 3,000 light-years away in the northern constellation Draco.

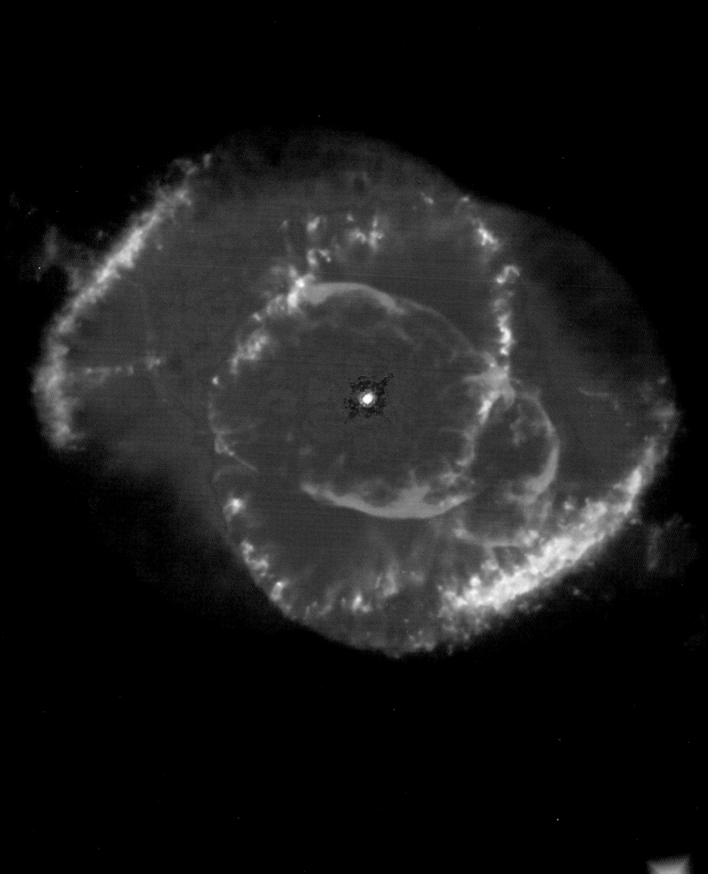

A star doing this is called a **PLANETARY NEBULA.** (Note that this term has nothing directly to do with planet formation; it is just a term carried down from the past when understanding was not as complete as it is today.)

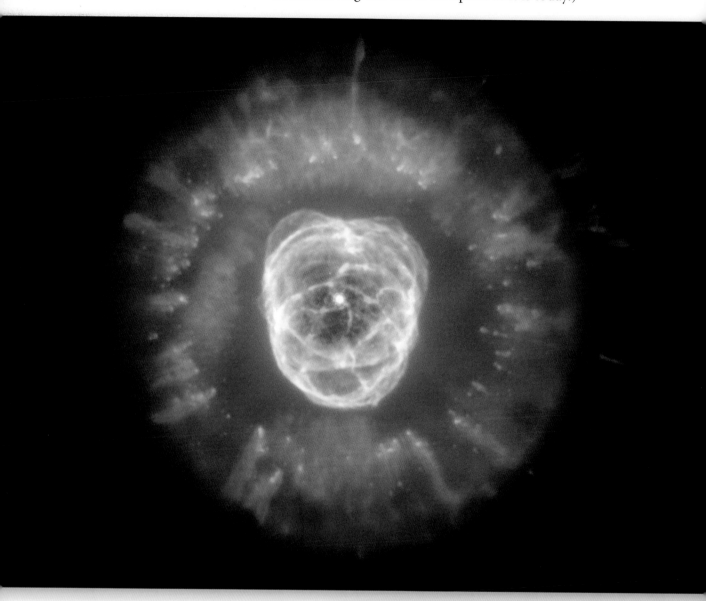

Eskimo Nebula

▲ This is a planetary nebula, the glowing remains of a dying, Sunlike star called the "Eskimo" Nebula (NGC 2392). The bright central region is a bubble of material being blown into space by the central star's intense "wind" of high-speed material. Scientists believe that a ring of dense material around the star's equator, ejected during its red giant phase, created the nebula's shape. The bubble is about 1 light-year long and about half a light-year wide. The Eskimo Nebula is about 5,000 light-years from Earth in the constellation Gemini. The nebula's glowing gases produce the colors in this image: nitrogen (red), hydrogen (green), oxygen (blue), and helium (violet).

White dwarfs and black dwarfs

When the gas and dust blow away from the planetary nebula, the outer part of the star disappears, leaving behind just the core, which is now called a **WHITE DWARF**.

A white dwarf star is a faint star that represents the remains of a much brighter star. It has a **MASS** about the same size as the Sun but a size about the same as the Earth, giving it a **DENSITY** about a 1,000 times that of water.

In general, the more massive a white dwarf is, the smaller it is. That is because the greater mass causes a greater **GRAVITY** to develop, and thus the squashing power of the star is higher. This intense gravity can only be counteracted by the gas left inside the star when it has become very dense. So, the bigger the mass, the smaller the star becomes before the gases can balance gravity.

In the core of a white dwarf is a mixture of carbon and oxygen, and around it is a shell of helium and hydrogen.

A white dwarf is only poorly luminous because its fuel is spent. So, what you see is the light sent out by the cooling star. No new **RADIATION** is being produced, and so the star will get fainter and fainter. Because of the tremendous store of heat and the fact that the outer shell helps insulate the core from heat loss, this does, however, take billions of years. But in the end the star stops sending out any radiation and is called a **BLACK DWARF**.

Because the amount of light sent out by a white dwarf is low compared with the amount of light at any other stage of the star's life, white dwarfs can only be seen if they are relatively close to us, perhaps less than a thousand light-years away. Black dwarfs do not shine and are almost invisible.

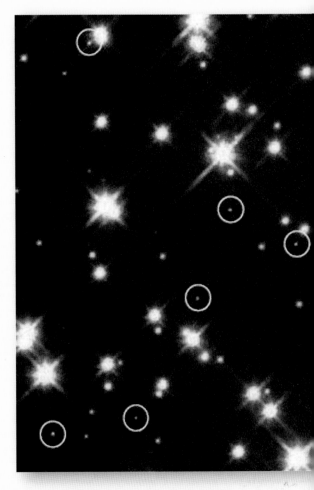

▲ White dwarf stars (ringed).

BLACK DWARF A degenerate star that has cooled so that it is now not visible.

DENSITY A measure of the amount of matter in a space.

GRAVITY The force of attraction between bodies.

MASS The amount of matter in an object.

PLANETARY NEBULA A compact ring or oval nebula that is made of material thrown out of a hot star.

RADIATION The transfer of energy in the form of waves (such as light and heat) or particles (such as from radioactive decay of a material).

WHITE DWARF Any star originally of low mass that has reached the end of its life.

Novas

As a pair of stars (binary stars) end their lives, the larger one may begin to swell up and leave the main sequence, while the smaller one remains part of the main sequence. When this happens, the gases in the envelope of the larger star spread out and are attracted by the **GRAVITATIONAL FIELD** of the smaller star. As a result, streams of gas flow from the larger to the smaller star until the once smaller star becomes the larger of the two. This makes the growing star leave the main sequence as well and become a giant star.

That causes an enormous release of light, and a faint star may be visible in the sky for the first time. That is why it is called a **NOVA**, from the Latin meaning "new." The situation then reverses, with material from the giant star flowing back to the smaller star, and so the brightness rapidly decreases. Notice that the stars return to their former brightness. In **SUPERNOVAS** (described on pages 31–33) this results in a huge explosion.

The death of massive stars

The sequence described above is what happens to a small star. During its final life stages a massive star (a Population I star in one of the arms of a **SPIRAL GALAXY**) can become so hot that it triggers nuclear reactions in the outer shell, and then the resulting death throws are very different.

The last stages of a massive star begin as a **RED GIANT**. But in a massive star more reactions, first in the core and then in the outer shell, can produce even heavier elements, such as silicon, sulfur, nitrogen, and calcium. Iron will eventually be formed in the core.

At this stage the process stops because an iron atom cannot be forced to release nuclear energy by adding more **PROTONS** or **NEUTRONS**. This is the planetary nebula phase (see pages 26–28).

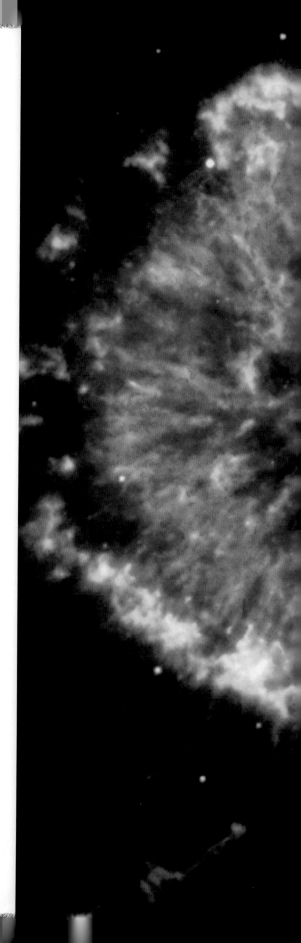

▶ This is the nebula NGC 6751 in the constellation Aquila. You can see the hot star in its center. It was responsible for ejecting the clouds of gas now seen encircling it.

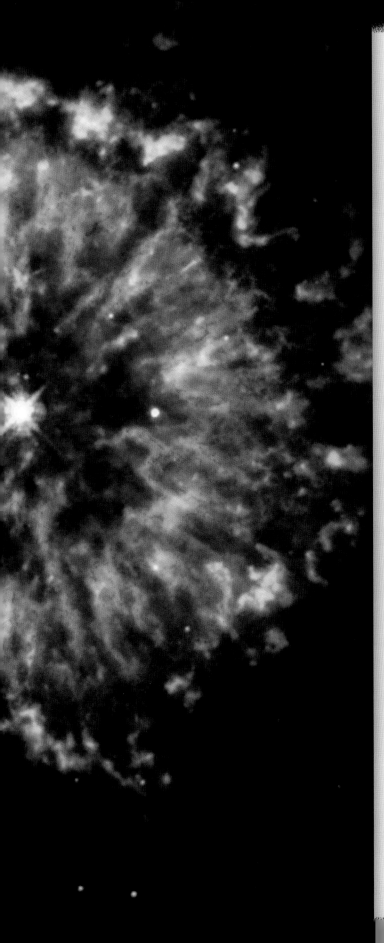

GRAVITATIONAL FIELD The region surrounding a body in which that body's gravitational force can be felt.

NEUTRINOS An uncharged fundamental particle that is thought to have no mass.

NEUTRONS Particles inside the core of an atom that are neutral (have no charge).

NEUTRON STAR A very dense star that consists only of tightly packed neutrons. It is the result of the collapse of a massive star.

NOVA (pl. **NOVAE**) A star that suddenly becomes much brighter, then fades away to its original brightness within a few months.

PROTONS Positively charged particles from the core of an atom.

RED GIANT A cool, large, bright star at least 25 times the diameter of our Sun.

SPIRAL GALAXY A galaxy that has a core of stars at the center of long curved arms made of even more stars arranged in a spiral shape.

SUPERNOVA A violently exploding star that becomes millions or even billions of times brighter than when it was younger and stable.

Supernovas

Iron only forms in the most massive of stars. When it does, it stops the reaction, and the core starts to cool rapidly. As a result, there is nothing to counteract the force of gravity. As the star nears the end of its planetary nebula phase, it collapses inward (implodes), creating a very small, extremely dense rotating core made almost entirely of neutrons—a **NEUTRON STAR**. At the same time, showers of **NEUTRINOS** are produced that add energy to the outer shell as it implodes and create an outward force that throws the shell into space to produce the most gigantic explosion in the universe.

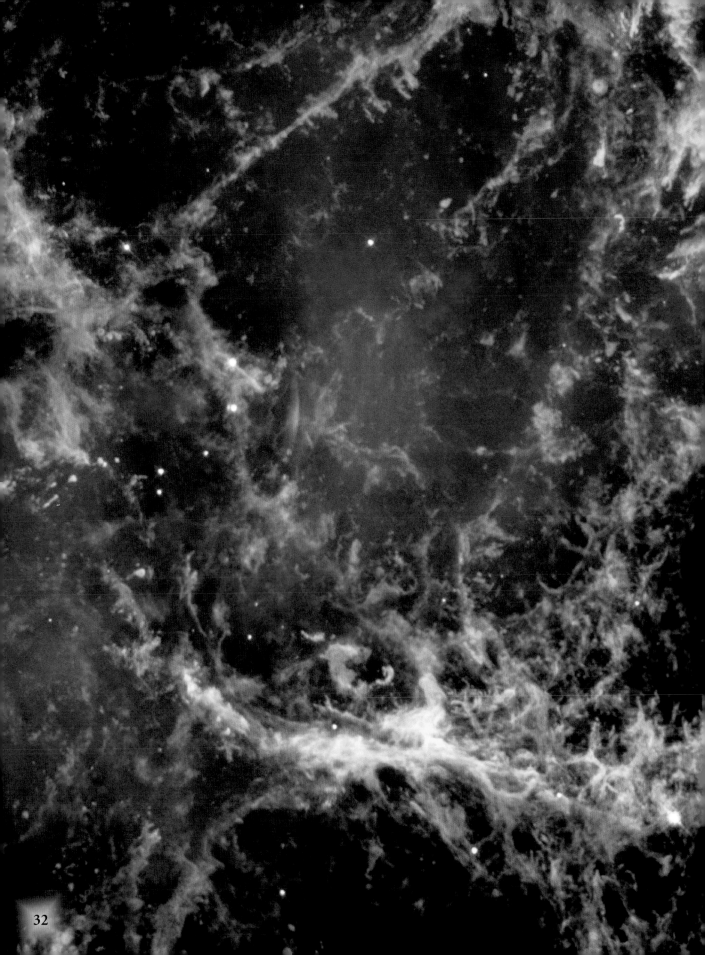

◄ This picture shows the Crab Nebula (catalogue numbers NGC 1952 and M1). It is one of the most-studied objects in the sky. It is about 5,000 light-years from the Earth and about 10 light-years across. It is the remnant of a violently exploding star, or **SUPERNOVA**. It was the first object listed (hence M1) in Charles Messier's catalogue of nebulous objects (see page 12).

In the year A.D. 1054 Chinese astronomers saw a new star appear. This star was so bright that it was visible in broad daylight for several weeks. Today the Crab Nebula is visible at the site of that bright star.

The Crab Nebula is still expanding after its explosion at the rate of 1,100 km per second. It contains an object called a pulsar (see page 34), which might be one feature of the collapsed supernova.

Because the gases in the expanding shell are extremely hot, they release light, and that is why they can be seen. It is from this cloud of elemental particles thrown into space that material is produced ready to form a new star such as our Sun.

Meanwhile, the **NEUTRON STAR** that forms in the center may be just 20 km across but may have a weight equal to several Suns. In this case it may be associated with a **PULSAR** (page 34) and a **QUASAR** (see page 35).

In some cases it may grow even bigger and denser. When this happens, it is so dense that its gravitational force is once more enormously high, and that turns the neutron star into a **BLACK HOLE** (see pages 35–39). The black hole will then start sucking in everything that comes within its reach.

A supernova event happens frighteningly fast. After millions of years of stability the supernova event happens in days, the brightness lasting for only a matter of weeks.

During this event the explosion makes the star shine more brightly than all of the rest of the stars in its galaxy put together.

The most famous supernova that has been recorded was observed in A.D. 1054. Its remains are now called the Crab Nebula. The Crab Nebula is different from a **PLANETARY NEBULA** because in its center there is a rapidly spinning, pulsating neutron star called a pulsar.

Because most of the stars in the universe are small, most of those that can be seen in their last stages are decaying to white dwarfs without any spectacular explosion to mark their dying days.

BLACK HOLE An object that has a gravitational pull so strong that nothing can escape from it.

NEUTRON STAR A very dense star that consists only of tightly packed neutrons. It is the result of the collapse of a massive star.

PLANETARY NEBULA A compact ring or oval nebula that is made of material thrown out of a hot star.

PULSAR A neutron star that is spinning around, releasing electromagnetic radiation, including radio waves.

QUASAR A rare starlike object of enormous brightness that gives out radio waves, which are thought to be released as material is sucked toward a black hole.

SUPERNOVA A violently exploding star that becomes millions or even billions of times brighter than when it was younger and stable.

▼ A neutron star (arrowed).

Pulsars (pulsating radio stars)

PULSARS emit very regular pulses of RADIO WAVES. They are associated with NEUTRON STARS.

When a neutron star forms during a supernova phase, NEUTRONS on the outer edge of the stars change into PROTONS and ELECTRONS, which have electrical charges. Because they are charged particles, they then get caught in the MAGNETIC FIELD surrounding the supernova and spin with it at speeds approaching that of light.

The only places where the particles can escape the magnetic field are at the magnetic poles, and that produces the point sources of radio waves that we can detect. The magnetic poles are not at the same places as the geographic poles, and so they move with the ROTATION of the star. (To visualize this, think of the Earth's north magnetic pole being at New York. As the Earth spins, the pole at New York faces out into different parts of space.)

Although the poles are continuously giving off radio waves, the waves are only sent in our direction when the spin of the star turns a pole to face us. This is similar to the way a lighthouse appears to be giving off flashing beams when in fact it is putting out a constant beam of light.

Pulsars have very short lives and may last no more than 10 million years, after which the magnetic fields are not strong enough to create the effect.

There is a pulsar in the Crab Nebula, which was probably formed at the same time as the supernova, about A.D. 1054 (see page 33).

▲ This picture shows a pulsar within the Crab Nebula, a rapidly rotating neutron star the size of Manhattan.

Bright wisps can be seen moving outward at half the speed of light to form an expanding ring that is visible in both X-ray and optical images.

Another dramatic feature is a turbulent jet that lies perpendicular to the inner and outer rings. It is a stream of matter moving at half the speed of light!

(See also pages 32–33.)

So far, over 300 pulsars are known that spin every 4 seconds or faster. One has been recorded spinning at 642 times per second. However, it is believed that there are one million active pulsars in the Milky Way Galaxy alone. To sustain this number, a pulsar, like its parent supernova, must be born every few decades.

Quasars

QUASAR stands for quasistellar radio sources. It represents a powerful source of radio waves, **X-RAYS**, and light.

Quasars are no more than a light-year or two in size, but they are up to 1,000 times brighter than giant galaxies. They are true pinpoints of light in the sky.

Quasars were first discovered in 1963. Since then, more than 15,000 quasars have been found, and we now know that they are the luminous cores at the heart of many galaxies. Quasars send out **RADIATION** a million million times as powerful as the radiation from the Sun from a space no bigger than the solar system.

Black holes

Neutron stars, pulsars, and quasars are all associated with very small sources of intense energy set within the remains of a star in a galaxy.

But they may all have something in common. They may be connected to another feature of some galaxies that we cannot see, and that does not directly send out radio or X-ray waves. We only know this feature is there because of its effect on its surroundings.

ELECTRONS Negatively charged particles that are parts of atoms.

MAGNETIC FIELD The region of influence of a magnetic body.

NEUTRONS Particles inside the core of an atom that are neutral (have no charge).

NEUTRON STAR A very dense star that consists only of tightly packed neutrons. It is the result of the collapse of a massive star.

PROTONS Positively charged particles from the core of an atom.

PULSAR A neutron star that is spinning around, releasing electromagnetic radiation, including radio waves.

QUASAR A rare starlike object of enormous brightness that gives out radio waves, which are thought to be released as material is sucked toward a black hole.

RADIATION The transfer of energy in the form of waves (such as light and heat) or particles (such as from radioactive decay of a material).

RADIO WAVES A form of electromagnetic radiation, like light and heat. Radio waves have a longer wavelength than light waves.

ROTATION Spinning around an axis.

X-RAY An invisible form of radiation that has extremely short wavelengths just beyond the ultraviolet.

◄ This is a quasar in the southern constellation Hydra, located about 10 billion light-years away. In this image we can only see things that are bigger than 10,000 light-years across.

The quasar is the pointlike object at the center of the images. It has two arclike and knotty tails extending in different directions. The tails are probably due to the closeness of the galaxy the quasar is in and to a nearby galaxy. The longer, southern tail extends over more than 150,000 light-years, one and a half times the diameter of the Milky Way.

▲ If you were on a planet near a black hole, this is what you might see: a small central source of light, perhaps a neutron star or quasar. Surrounding it there would be a disk—called an accretion (buildup) disk—that contains material sucked in from the surrounding space. Coming out at right angles to the disk like a searchlight is a stream of electrons.

BINARY STAR A pair of stars that are gravitationally attracted, and that revolve around one another.

BLACK HOLE An object that has a gravitational pull so strong that nothing can escape from it.

GRAVITATIONAL FIELD The region surrounding a body in which that body's gravitational force can be felt.

PHOTON A particle (quantum) of electromagnetic radiation.

We cannot detect the feature because it has **GRAVITATIONAL FIELDS** so massive it sucks in everything, and even light (**PHOTONS** of radiation) cannot escape it. It is known as a **BLACK HOLE** (and also as an active galactic nucleus).

Although the term black hole suggests a tunnel rather than a physical object, a black hole is not an absence of matter. It is the most dense form of matter imaginable, with the most extreme gravitational field in the universe. If the Earth were to become a black hole, it would measure no more than a few millimeters across.

The gravitational field is what sucks dust, gas, and other stars toward it, and what stops light from leaving it, which is why we cannot see it.

A black hole might be 2 light-hours across (the distance from the Sun to the Earth is 8 light-minutes). It might have a mass of a hundred million or a billion Suns.

To keep such a phenomenal gravitational field intact, the black hole has to add continuously to its mass. A black hole does so by devouring gas, dust, and even stars that happen to come too close to it. The mass each black hole needs is the equivalent of several Suns every year.

Why we can "see" black holes

Although nothing flows out of a black hole, as material flows toward it, the material heats up, and it gives out light rays, X-rays, and radio waves. Black holes can also be observed by the effects of their enormous gravitational fields on nearby objects in space.

For example, if a black hole develops from a massive star that was part of a **BINARY STAR** (twin) system, it will begin to devour its companion star. As this happens, it will cause the companion star to heat up and send out X-rays. The binary system Cygnus X-1 is made of a blue supergiant and an invisible companion star—the black hole.

The black hole at the center of the M87 galaxy has a mass equal to two to three billion Suns, and it affects the gases near to it by causing them to swirl around, like water draining from a bath. So "radio stars" (see page 45) may be the signature of material flowing toward black holes.

◄ This is the black hole-powered core of a nearby active galaxy 13 million light-years away in the southern constellation Circinus. These active galactic nuclei have the ability to remove gas from the centers of their galaxies by blowing it out into space at phenomenal speeds.

At the center of the starburst the black hole and its accretion disk are expelling gas out of the galaxy's disk and into its halo (the region above and below the disk). The detailed structure of this gas is seen as magenta-colored streamers extending toward the top of the image.

▼ Black holes are not far away. This diagram shows the path of a black hole believed to be within the Milky Way Galaxy. The Sun's orbit is indicated for comparison, showing that at some time the orbits may well come quite close. The black hole is moving at 400,000 kilometers an hour.

For example, the radio source called Sagittarius A*, some 30,000 light-years from us, is thought to mark a black hole.

The same process powers the jets of charged particles flung from the black hole, which we can see as blue streaks in space and also the shining core that we observe as a pointlike QUASAR.

Black holes are not uncommon. The closest black hole to us is within the central bulge of the MILKY WAY. A black hole is a natural part of a star's formation, the star and the black hole being created as part of the life cycle of a star. Evidence for this comes from

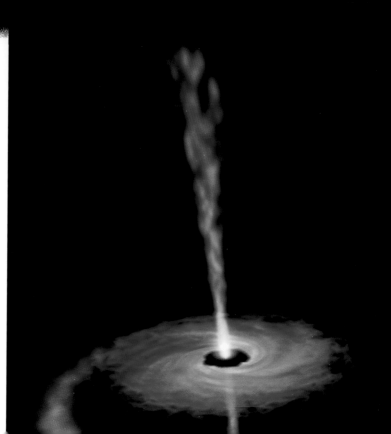

◀ Streaming out from the center of the elliptical (oval) galaxy M87 in the Virgo cluster, like a cosmic searchlight, is one of nature's most amazing phenomena, a black hole-powered jet of electrons and other subatomic particles traveling at nearly the speed of light. The blue of the jet contrasts with the yellow glow from the combined light of billions of unseen stars and the yellow, pointlike globular clusters that make up this galaxy.

Lying at the center of M87 is a supermassive black hole, which has swallowed up a mass equivalent to two billion times the mass of our Sun. The jet originates in the disk of superheated gas swirling around this black hole and is propelled and concentrated by the intense, twisted **MAGNETIC FIELDS** trapped within this **PLASMA**. The light that we see (and the radio emission) are produced by electrons twisting along magnetic field lines in the jet, a process known as synchrotron radiation, which gives the jet its bluish tint.

MAGNETIC FIELD The region of influence of a magnetic body.

MILKY WAY The spiral galaxy in which our star and solar system are situated.

PLASMA A collection of charged particles that behaves something like a gas. It can conduct an electric charge and be affected by magnetic fields.

QUASAR A rare starlike object of enormous brightness that gives out radio waves, which are thought to be released as material is sucked toward a black hole.

◀ An artist's impression of material from a sun being spun off into a black hole. The blue disk is called the accretion disk. The black hole is at its center. Powerful jets of radiation are given off at right angles to the black hole.

3: GALAXIES

When you look up at the night sky, you might think that there are **STARS** everywhere. But stars are not scattered evenly. Instead, they are clumped into certain regions with very little **MATTER** between them. These concentrated regions of matter and stars are **GALAXIES**.

Galaxies are also the places in the **UNIVERSE** where matter is transformed into energy. They are, then, both the source and location of everything. They are the factories of the universe.

Galaxies contain matter that will emit light if it gets hot enough or will **REFLECT** light from other bodies. However, it is certain that what we see is not all there is. Calculations show that a galaxy has more **MASS** in it than we can see, so the rest must be made up of matter that is invisible to us—that is, it gives out no light. This is called **DARK MATTER**. It, too, can be a source of energy, although as yet we know little about it.

The nature of galaxies

Our galaxy, which we call the **MILKY WAY**, is also known by astronomers as *the* galaxy. All galaxies contain a huge mass of gas, dust, and stars spinning slowly around a center. Many have huge trailing spiral arms. At the center of such a galaxy is a bulging region made of stars. Some galaxies also have a flat disk made of stars that are less concentrated than in the bulge, but that form spiral arms and a halo beyond the spiral arms or main disk made of old stars.

Space gas and dust—interstellar (between the stars) matter—are found everywhere in a galaxy, although they make up only about 1% of the mass of a galaxy. Ninety-nine percent of the mass of a galaxy is concentrated in the stars.

▶ The Andromeda Galaxy (catalogues NGC 224 and M31) is a typical **SPIRAL GALAXY** in the constellation Andromeda. It is the separate galaxy nearest the Milky Way Galaxy. It can be seen in the night sky without a telescope. It is 2,000,000 light-years from the Earth. It is 200,000 light-years across.

DARK MATTER Matter that does not shine or reflect light.

GALAXY A system of stars and interstellar matter within the universe.

MASS The amount of matter in an object.

MATTER Anything that exists in physical form.

MILKY WAY The spiral galaxy in which our star and solar system are situated.

REFLECT To bounce back any light that falls on a surface.

SPIRAL GALAXY A galaxy that has a core of stars at the center of long curved arms made of even more stars arranged in a spiral shape.

STAR A large ball of gases that radiates light. The star nearest the Earth is the Sun.

UNIVERSE The entirety of everything there is; the cosmos.

The Sun (with its planets that make up the solar system) is one of these stars, lying near the edge of one of the spiral arms within the flat disk of our galaxy.

How stars move in a galaxy

The stars are not stationary in a galaxy but revolve around it. The time they take to move around the center is extremely long—something like 200 million years. The fact that stars rotate around the center of the galaxy shows that they are bound inside it just as the planets are bound inside the solar system. It shows that they are held together by the universal force of GRAVITY.

How galaxies were formed

Because galaxies are the main features of the COSMOS, understanding how they formed is key to understanding much about how the cosmos works. Yet, so fundamental is this question and so remote the source of information, that the origin of galaxies is still shrouded in uncertainty.

One suggestion is that each galaxy must have begun through the rapid collapse of a single large gas cloud. In this cloud some stars emerged quickly, perhaps in the first 100 million years, while others formed more slowly.

An alternative suggestion is that a galaxy builds slowly by the clumping together of material.

Stars produce light when hydrogen burns to helium (see page 23). Helium now makes up about a quarter of the mass of our galaxy. So, it is possible to figure out the amount of stellar (star) matter needed to produce it. But that turns out to be much too small to account for all of the helium in our galaxy. So, although helium is produced by stars, most of the helium was not made in stars and must have been made much earlier.

Edwin Powell Hubble (1889–1953)

Hubble is recognized as one of the great 20th-century astronomers and one of the first to think about an expanding universe.

His observations at Mount Wilson Observatory provided the first evidence that NEBULAE exist beyond the Milky Way.

Hubble began to classify galaxies into spirals, ellipticals, and irregulars (see right). He proved that galaxies were more or less evenly distributed in space for as far as could be seen. This supported the idea, known as the COSMOLOGICAL PRINCIPLE, that the universe is uniform.

It was while making this classification that Hubble discovered that all galaxies are moving away from one another, that is, that the universe is expanding.

He then found that the universe was expanding in such a way that the ratio of the speed of the galaxies to their distance is a constant. This idea is known as Hubble's Law, and the constant became known as Hubble's constant, H. It is a cornerstone of modern ASTRONOMY (see page 53).

The value of H is between 15 and 30 km per second per 1,000,000 light-years. By using this constant, it is possible to take the inverse (reciprocal) and deduce that the time since the start of the universe (see page 54) is 10 billion to 15 billion years.

Type 1: Irregular galaxies

◄ Irregular galaxy (NGC1705).

ASTRONOMY The study of space beyond the Earth and its contents.

COSMOLOGICAL PRINCIPLE States that the way you see the universe is independent of the place where you are (your location). In effect, it means that the universe is roughly uniform throughout.

COSMOS The universe and everything in it. The word "cosmos" suggests that the universe operates according to orderly principles.

GRAVITY The force of attraction between bodies.

NEBULA (pl. **NEBULAE**) Clouds of gas and dust that exist in the space between stars.

Type 2: Elliptical galaxies

▶ (NGC 4261).

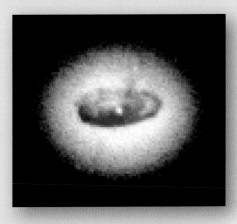

Type 3: Spiral galaxies
Subtype: Normal spiral galaxies

▼ Normal spiral (NGC 4622).

Type 3: Spiral galaxies
Subtype: Barred spiral galaxies

▼ Barred spiral (NGC 2787).

Types of galaxies

One important step to finding out about galaxies is to look for patterns among them. Once you have done this, you can group galaxies into similar types. This is called classification.

Edwin Hubble was the first person to classify galaxies (pages 42–43). The first type he called "irregulars," having no apparent pattern of bright stars inside them. The light from them tends to be blue-white, which is an indication of young, hot stars. It is likely that these galaxies are creating huge numbers of stars, but that their development has not yet reached a stage where they show any organization. Our neighboring galaxies, the Large and Small Magellanic Clouds (see page 51), are irregular galaxies of this kind.

Besides the irregular galaxies there are ELLIPTICAL GALAXIES and SPIRAL GALAXIES. Elliptical galaxies tend to be football shaped (oval). Spiral galaxies are flat and disk shaped. Some spiral galaxies have curved arms; others have bars of stars and are known as barred spirals.

The ellipticals and spirals account for most of the galaxies in the universe.

Radio galaxies

Although stars are all amazingly good at converting large amounts of matter into energy, the most powerful ones do not necessarily convert matter into light energy. When scientists study the universe in radio or X-ray wavelengths rather than in visible light, they find it studded with galaxies that are completely invisible to an ordinary telescope. They are called RADIO GALAXIES and were once referred to as radio stars (see page 36).

ELLIPTICAL GALAXY A galaxy that has an oval shape rather like a football, and that has no spiral arms.

RADIO GALAXY A galaxy that gives out radio waves of enormous power.

SPIRAL GALAXY A galaxy that has a core of stars at the center of long curved arms made of even more stars arranged in a spiral shape.

▲ This is the galaxy Centaurus A (NGC 5128) in the southern sky. Its appearance is due to an opaque dust lane that covers the central part of the galaxy. This dust is likely to be the remains of a cosmic merger between a giant elliptical galaxy and a smaller spiral galaxy full of dust.

Centaurus A is one of the brightest radio sources in the sky. It is 11 million light-years away and is also the nearest radio galaxy. The radio emission is probably due to material being sucked into a massive black hole.

◄ This is the Whirlpool Galaxy (catalogue numbers NGC 5194 and M51). It can be seen with smaller telescopes.

You can see numerous shining clusters of young and energetic stars. The bright clusters are surrounded by glowing hydrogen gas.

Along the spiral arms dust "spurs" are seen branching out almost perpendicular to the main spiral arms. There is a dust disk in the nucleus, which may provide fuel for a black hole.

4: THE UNIVERSE

We began this book with a brief introduction to the **UNIVERSE**. It gave us a framework for understanding the **STARS** and **GALAXIES** that make up the universe. Now we can look in more detail at what the universe itself is like.

Asking what the universe is like and how it came to be that way is the most fundamental of questions. We still do not have an answer to match all of our observations. Perhaps we never will. In fact, the more we find out about **SPACE**, the more curious it becomes.

New rules

There is a fundamental problem. When you look over the vast distances and huge time spans that exist in the universe, with objects moving very fast, the laws, or rules, that we use on Earth do not hold true.

The laws we have developed are approximations of the truth and are good enough for use on Earth. But, as Albert Einstein showed with his **THEORY OF RELATIVITY** (see page 48), it is a different matter when you look out into space.

Einstein was concerned with **GRAVITY**. He showed that **PHOTONS** of light, even though they always apparently traveled in straight lines, would return to where they started from without ever coming to an edge or boundary. It is gravity that produces this effect. The fact that gravity can bend light locally can even be seen in many observations of the way light moves when it is close to massive stars.

This new set of rules is called relativity. Using the arguments of relativity, it is possible to think of almost unimaginable things, such as a universe that has no edge and space that is curved.

GALAXY A system of stars and interstellar matter within the universe.

GRAVITATIONAL FIELD The region surrounding a body in which that body's gravitational force can be felt.

GRAVITY The force of attraction between bodies.

LIGHT-YEAR The distance traveled by light through space in one Earth year, or 63,240 astronomical units.

PHOTON A particle (quantum) of electromagnetic radiation.

SPACE Everything beyond the Earth's atmosphere.

STAR A large ball of gases that radiates light. The star nearest the Earth is the Sun.

THEORY OF RELATIVITY A theory based on how physical laws change when an observer is moving. Its most famous equation says that at the speed of light, energy is related to mass and the speed of light.

UNIVERSE The entirety of everything there is; the cosmos.

◄ Wherever you look in the sky with a powerful telescope, you will find galaxies. This is a massive cluster of galaxies called Abell 2218. It is in the constellation Draco, some 2 billion **LIGHT-YEARS** from Earth.

The cluster is so massive that its enormous **GRAVITATIONAL FIELD** bends light rays passing through it, much as an optical lens bends light to form an image. This phenomenon magnifies, brightens, and distorts images from faraway objects just as a powerful "zoom lens" does. As a result, we can see distant galaxies that could not normally be observed with the largest telescopes. In this picture you can see spiral and elliptical galaxies. The yellow-white color of several of the galaxies represents the combined light of many stars.

Albert Einstein and the theory of relativity

Albert Einstein (1879–1955) is regarded as one of the most important figures of 20th-century science. He developed a **THEORY OF RELATIVITY** in 1905, in which he showed that there was a special relationship between space and time.

Although the effect of this relationship is not apparent in our everyday lives, it becomes vitally important when thinking about the nature of space because of the vast distances, huge gravities, and great speeds involved.

Einstein showed that it was possible for energy to be transformed into matter and vice versa. The well-known formula for this is:

$$e = mc^2$$

where e = energy

m = mass

c = the speed of light

By 1916 Einstein had developed his theory further, showing in his theory of general relativity that gravity is also related to space and time.

◄ Although it is easy to see the formation of galaxies and changes to stars, it seems far more difficult to identify stars with planetary systems in the making. This may just be what can be seen in the star DG Tau and in the dust disk surrounding it. It is a young star ringed by a swirling disk that may spin off planets. DG Tau has not yet begun to burn hydrogen in its core. Such stars are called T-Tauri objects (see page 18).

Since 1995 astronomers have detected more than 100 extrasolar planets, many considered too large and close to their hot parent stars to sustain life. One in four lies within 16 million kilometers of the parent star. There is a gap of nearly 29 million kilometers between DG Tau and its orbiting dust disk.

▲ We can see the shape of this galaxy because it is far from us, and we are not part of it.

Another new rule that we need when dealing with space is called the **COSMOLOGICAL PRINCIPLE**. It states that wherever you are in the universe, and in whatever direction you look, the universe seems much the same. Its importance is that it suggests that the universe has no edge.

The universe has a structure

Let's begin by looking for some order in the universe. In this book we have seen how stars form parts of large structures called galaxies.

COSMOLOGICAL PRINCIPLE States that the way you see the universe is independent of the place where you are (your location). In effect, it means that the universe is roughly uniform throughout.

THEORY OF RELATIVITY A theory based on how physical laws change when an observer is moving. Its most famous equation says that at the speed of light, energy is related to mass and the speed of light.

We are part of the solar system. The Earth moves around the star we called the Sun. The Sun is on the outer part of an arm of a moderate-sized galaxy that we call the **MILKY WAY**.

If we use powerful telescopes, we can see that there are other galaxies. The more we look at stars and clusters of stars, the more we see them as parts of an even greater group. Some are far away, others much closer to us.

The Local Group

Astronomers now call the galaxies that are our near neighbors the **LOCAL GROUP**. They include our galaxy, the Milky Way, the neighboring smaller **SATELLITE** galaxies of the **MAGELLANIC CLOUDS**, the Andromeda Galaxy, our nearest independent galaxy, and over 20 other relatively near galaxies. The Milky Way and Andromeda are the biggest members of the Local Group.

Our galaxy is made up of a number of arms that spiral to the central hub. This is called a **SPIRAL GALAXY** (see pages 42–45). The Andromeda Galaxy is another spiral galaxy.

Looking farther afield, we see that many galaxies in our Local Group are not, in fact, spirals but form great football-shaped masses called **ELLIPTICAL GALAXIES**. They are bigger and brighter, and are not flattened into a disk like the spirals.

Groups of galaxies, like our Local Group, cluster together throughout the cosmos, but many contain vastly more galaxies, often numbering thousands rather than the few that make up the Local Group. But on a large scale the size range and spacing of galaxies are much the same in all directions.

Light as a time machine

We have already said that we use the speed of light as a measure for space. Some galaxies are several billion light-years from the Earth. Light reaching us from these galaxies is so old that it left the galaxy at just about the time that the first simple life appeared on the Earth.

ELLIPTICAL GALAXY A galaxy that has an oval shape rather like a football, and that has no spiral arms.

LOCAL GROUP The Milky Way, the Magellanic Clouds, the Andromeda Galaxy, and over 20 other relatively near galaxies.

MAGELLANIC CLOUD Either of two small galaxies that are companions to the Milky Way Galaxy.

MILKY WAY The spiral galaxy in which our star and solar system are situated.

NEBULA (pl. **NEBULAE**) Clouds of gas and dust that exist in the space between stars.

RADIATION The transfer of energy in the form of waves (such as light and heat) or particles (such as from radioactive decay of a material).

SATELLITE An object that is in an orbit around another object.

SPIRAL GALAXY A galaxy that has a core of stars at the center of long curved arms made of even more stars arranged in a spiral shape.

STELLAR WIND The flow of tiny charged particles (called plasma) outward from a star.

ULTRAVIOLET A form of radiation that is just beyond the violet end of the visible spectrum and so is called "ultra" (more than) violet. At the other end of the visible spectrum is "infra" (less than) red.

Stars in the Magellanic Clouds

The Magellanic Clouds are two irregular galaxies, or NEBULAE, just outside the Milky Way Galaxy. They were first recorded by the Portuguese navigator Ferdinand Magellan during the first voyage around the world. They can be seen with the unaided eye in the southern hemisphere.

The Large Magellanic Cloud (LMC) is 150,000 light-years from the Earth, and the Small Magellanic Cloud (SMC) is 200,000 light-years away.

The Magellanic Clouds are younger than the Milky Way. They contain many young stars and star clusters.

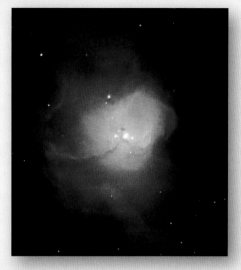

▲ This is a picture of the self-destruction of a massive star called Supernova 1987A in the Large Magellanic Cloud. The explosion was observed first on February 23, 1987, although it took place 150,000 years ago, because the star is 150,000 light-years away. This picture shows the supernova remnant, surrounded by inner and outer rings of material. The many bright blue stars near the supernova are massive stars, each more than six times more massive than our Sun.

◄ This picture shows the Small Magellanic Cloud with young, ultrabright stars nested in their cloud of glowing gases. There are 50 separate stars tightly packed in the core. These young stars are sending out huge STELLAR WINDS that are buffeting the walls of the nebula. The pair of bright stars in the center of the nebula is sending out the ULTRAVIOLET RADIATION that makes the nebula glow.

This is a true-color view.

Light coming from places that are more than 10 billion light-years away has not yet had time to travel to the Earth. That is why we talk about the "observable" universe. We know only about the billions of galaxies within this distance. It is certain that there are many, many more.

One interesting result of this is that light from distant galaxies provides a natural "time machine" for seeing far back in time.

The more distant the object, the longer it has taken for light to get from it to us, and therefore the nearer the start of the universe that light was emitted. So, this light gives us a snapshot of the early universe, not a snapshot of what it is like now. To know what it is like now, we have to assume that everywhere is similar to conditions in our neighborhood of space.

An expanding universe

When we start to investigate distant objects, we notice something curious. Every time a remote galaxy is measured, its distance from us is greater than when it was measured before.

No matter in which direction we look, the same is true. In effect, everything we see in the universe is speeding away from every other thing, just like points on the surface of an inflating balloon. So, we cannot escape the dramatic conclusion that in the past everything was much closer. The implications of this were first deduced by Edwin Powell Hubble in the form of Hubble's Law (see page 42). It is as though we are all part of some gigantic explosion that happened in the past.

◀ This is spiral galaxy NGC 4414. It was observed on 13 different occasions over the course of 2 months to accurately measure the distance to it, which was found to be 19.1 megaparsecs, or about 60 million light-years. This measurement contributes to astronomers' overall knowledge of the rate of expansion of the universe.

The Big Bang

If we trace the paths of all of the galaxies back in time, we find that they all point back to a single place of origin at a single point in time. This point existed 10,000,000,000 years ago. For us this is the start of the universe. This starting point is commonly called the **BIG BANG**.

Although we can see to a beginning in the Big Bang, the fact that there is a date for "creation" also raises questions we cannot answer. For us the Big Bang is the beginning—we cannot see beyond it. Yet we sense that there was something before the Big Bang in order for it to have happened at all. This is a question to which no one has an answer. So, as in the past, the more we discover, the more puzzling the universe becomes.

If the observable universe did begin with a Big Bang, what we see now is the debris of an incredible explosion that took place at that time. Every star, every planet, every **METEORITE**, every **ATOM** is part of that explosion.

The first moments of the universe

If the Big Bang idea is correct, then the universe began at a single point of unimaginably high temperature and high **DENSITY**.

Within a small fraction of a second of this beginning the universe was expanding, causing it to become less dense and cooler. Within a few seconds charged particles started to form, as energy was converted to matter, and the building blocks of **ELEMENTS** such as hydrogen and helium started to form, together with tiny uncharged particles called **NEUTRINOS**. These particles were not atoms. It took another thousand years before the universe was cool enough for them to form.

The early universe thus began mainly as **RADIATION** (energy). This radiation was able to continue to move through space, which is still does today, filling it with what is called **MICROWAVE RADIATION**.

▶ How the universe might have changed from the beginning, when there was just radiation, through inflation (expanding universe) to the formation of galaxies as we see today.

ATOM The smallest particle of an element.

BIG BANG The theory that the universe as we know it started from a single point (called a singularity) and then exploded outward. It is still expanding today.

DENSITY A measure of the amount of matter in a space.

ELEMENT A substance that cannot be decomposed into simpler substances by chemical means.

METEORITE A meteor that reaches the Earth's surface.

MICROWAVE RADIATION The background radiation that is found everywhere in space, and whose existence is used to support the Big Bang theory.

NEUTRINOS An uncharged fundamental particle that is thought to have no mass.

RADIATION The transfer of energy in the form of waves (such as light and heat) or particles (such as from radioactive decay of a material).

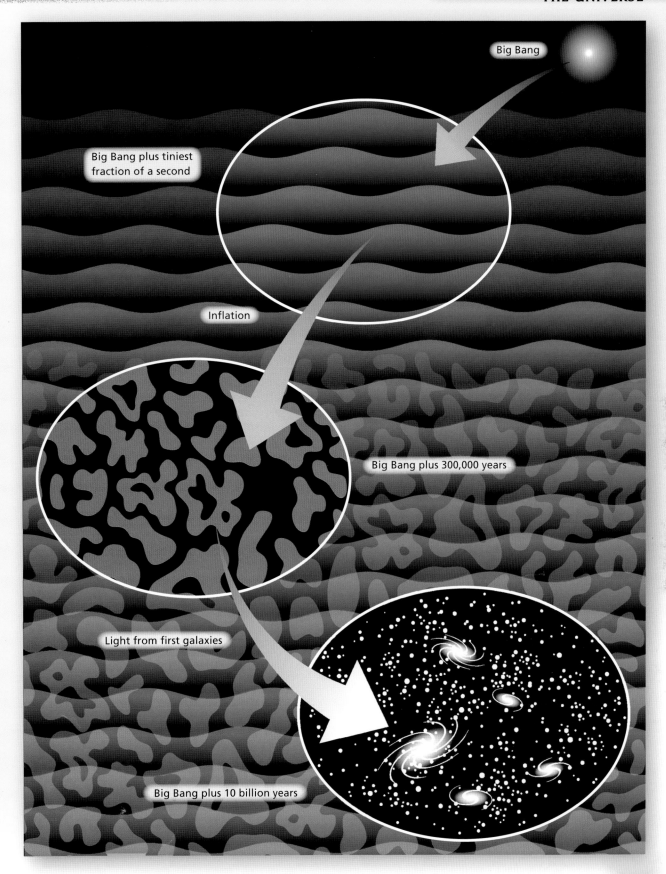

Big Bang

Big Bang plus tiniest
fraction of a second

Inflation

Big Bang plus 300,000 years

Light from first galaxies

Big Bang plus 10 billion years

Once atoms began to form, they could come together to form **MOLECULES**. **GRAVITY** began to affect the molecules scattered around as gas, drawing them together and forming stars and galaxies, and producing the universe as we know it today.

How can we support this idea? One way is to look for traces of the first Big Bang. Any explosion causes shock waves. Some are sound waves, some are light waves, and some are other kinds of radiation. These waves remain when the explosion seems to be long over.

Everywhere astronomers look today they find **MICROWAVE RADIATION**. This is difficult to explain because it is in every part of the universe. But one explanation for it would be that it is the remains of a once fierce light that accompanied the explosion, and that has now decayed to weak microwave radiation.

▼ This is a picture of the microwave background radiation for the entire sky as seen from the Earth. Reds show more, blues show less radiation. The fact that the pattern of reds and blues is even across the sky is evidence for microwave background radiation being everywhere. This is the main support for the Big Bang theory.

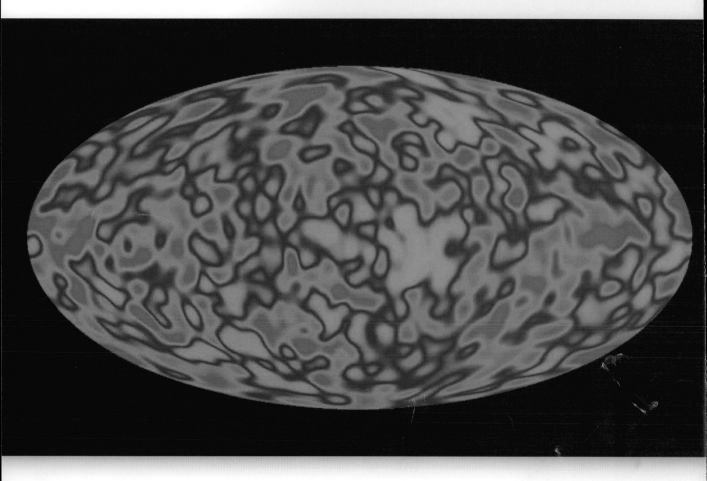

56

◄ Our star has planets surrounding it. It is likely that this is a commonplace feature of the universe. But planets are so small that they have never been seen before outside our galaxy. This picture shows a newborn star with a long band of gas (nebula) pointing toward a faint companion object (bottom left), which could be the first planet from another star system to be imaged directly. It may be a hot, young protoplanet several times the mass of Jupiter, orbiting at a distance of 210 billion kilometers from the star.

How the universe might continue

It is no easier to predict what might happen in the future than to wonder how it all began. We know that the universe is expanding, but it may or may not go on expanding forever. That depends on the mass of material in the universe, an amount that is still a matter of speculation. Some think the universe will continue to expand forever. Others believe that the universe will continue to expand for a while, then begin to collapse back to the point it came from.

It is likely that our views will change as we learn more about the universe. For example, we are only just finding examples of planets in other star systems. What else might there be to find?

Central to all of the wonder about the universe is whether space and time can have a beginning and an end (that is, they are finite), or whether there is no beginning and no end (that is, they are infinite). Will everything end when matter comes back together? Will there be a Big Crunch? We still do not know.

Planet discoveries

Recent sweeps of the domelike bulge of the Milky Way by the Hubble Space Telescope have discovered large numbers of new planets orbiting stars in our galaxy. As a result, in 2004 the number of planets orbiting stars in our galaxy has risen to 230 and is certain to increase. The discovery lends support to the idea that almost every sunlike star in our galaxy, and probably the universe, is accompanied by planets.

GRAVITY The force of attraction between bodies.

MICROWAVE RADIATION The background radiation that is found everywhere in space, and whose existence is used to support the Big Bang theory.

MOLECULE A group of two or more atoms held together by chemical bonds.

SET GLOSSARY

ABSOLUTE ZERO The coldest possible temperature, defined as 0 K or −273°C.
See also: **K**.

ACCELERATE To gain speed.

AERODYNAMIC A shape offering as little resistance to the air as possible.

AIR RESISTANCE The frictional drag that an object creates as it moves rapidly through the air.

AMINO ACIDS Simple organic molecules that can be building blocks for living things.

ANNULAR Ringlike.
An annular eclipse occurs when the dark disk of the Moon does not completely obscure the Sun.

ANTENNA (pl. **ANTENNAE**) A device, often in the shape of a rod or wire, used for sending out and receiving radio waves.

ANTICLINE An arching fold of rock layers where the rocks slope down from the crest.

ANTICYCLONE A roughly circular region of the atmosphere that is spiraling outward and downward.

APOGEE The point on an orbit where the orbiting object is at its farthest from the object it is orbiting.

APOLLO The program developed in the United States by NASA to get people to the Moon's surface and back safely.

ARRAY A regular group or arrangement.

ASH Fragments of lava that have cooled and solidified between when they leave a volcano and when they fall to the surface.

ASTEROID Any of the many small objects within the solar system.
Asteroids are rocky or metallic and are conventionally described as significant bodies with a diameter smaller than 1,000 km. Asteroids mainly occupy a belt between Mars and Jupiter (asteroid belt).

ASTEROID BELT The collection of asteroids that orbit the Sun between the orbits of Mars and Jupiter.

ASTHENOSPHERE The region below the lithosphere, and therefore part of the upper mantle, in which some material may be molten.

ASTRONOMICAL UNIT (**AU**) The average distance from the Earth to the Sun (149,597,870 km).

ASTRONOMY The study of space beyond the Earth and its contents. It includes those phenomena that affect the Earth but that originate in space, such as meteorites and aurora.

ASTROPHYSICS The study of physics in space, what other stars, galaxies, and planets are like, and the physical laws that govern them.

ASYNCHRONOUS Not connected in time or pace.

ATMOSPHERE The envelope of gases that surrounds the Earth and other bodies in the universe.
The Earth's atmosphere is very different from that of other planets, being, for example, far lower in hydrogen and helium than the gas giants and lower in carbon dioxide than Venus, but richer in oxygen than all the others.

ATMOSPHERIC PRESSURE The pressure on the gases in the atmosphere caused by gravity pulling them toward the center of a celestial body.

ATOM The smallest particle of an element.

ATOMIC MASS UNIT A measure of the mass of an atom or molecule.
An atomic mass unit equals one-twelfth of the mass of an atom of carbon-12.

ATOMIC WEAPONS Weapons that rely on the violent explosive force achieved when radioactive materials are made to go into an uncontrollable chain reaction.

ATOMIC WEIGHT The ratio of the average mass of a chemical element's atoms to carbon-12.

AURORA A region of illumination, often in the form of a wavy curtain, high in the atmosphere of a planet.
It is the result of the interaction of the planet's magnetic field with the particles in the solar wind. High-energy electrons from the solar wind race along the planet's magnetic field into the upper atmosphere. The electrons excite atmospheric gases, making them glow.

AXIS (pl. **AXES**) The line around which a body spins.
The Earth spins around an axis through its north and south geographic poles.

BALLISTIC MISSILE A rocket that is guided up in a high arching path; then the fuel supply is cut, and it is allowed to fall to the ground.

BASIN A large depression in the ground (bigger than a crater).

BIG BANG The theory that the universe as we know it started from a single point (called a singularity) and then exploded outward. It is still expanding today.

BINARY STAR A pair of stars that are gravitationally attracted, and that revolve around one another.

BLACK DWARF A degenerate star that has cooled so that it is now not visible.

BLACK HOLE An object that has a gravitational pull so strong that nothing can escape from it.
A black hole may have a mass equal to thousands of stars or more.

BLUE GIANT A young, extremely bright and hot star of very large mass that has used up all its hydrogen and is no longer in the main sequence. When a blue giant ages, it becomes a red giant.

BOILING POINT The change of state of a substance in which a liquid rapidly turns into a gas without a change in temperature.

BOOSTER POD A form of housing that stands outside the main body of the launcher.

CALDERA A large pit in the top of a volcano produced when the top of the volcano explodes and collapses in on itself.

CAPSULE A small pressurized space vehicle.

CATALYST A substance that speeds up a chemical reaction but that is itself unchanged.

CELESTIAL Relating to the sky above, the "heavens."

CENTER OF GRAVITY The point at which all of the mass of an object can be balanced.

CENTRIFUGAL FORCE A force that acts on an orbiting or spinning body, tending to oppose gravity and move away from the center of rotation.
For orbiting objects the centrifugal force acts in the opposite direction from gravity. When satellites orbit the Earth, the centrifugal force balances out the force of gravity.

CENTRIFUGE An instrument for spinning small samples very rapidly.

CHAIN REACTION A sequence of related events with one event triggering the next.

CHASM A deep, narrow trench.

CHROMOSPHERE The shell of gases that makes up part of the atmosphere of a star and lies between the photosphere and the corona.

CIRCUMFERENCE The distance around the edge of a circle or sphere.

COMA The blurred image caused by light bouncing from a collection of dust and ice particles escaping from the nucleus of a comet.

The coma changes the appearance of a comet from a point source of reflective light to a blurry object with a tail.

COMBUSTION CHAMBER A vessel inside an engine or motor where the fuel components mix and are set on fire, that is, they are burned (combusted).

COMET A small object, often described as being like a dirty snowball, that appears to be very bright in the night sky and has a long tail when it approaches the Sun.

Comets are thought to be some of the oldest objects in the solar system.

COMPLEMENTARY COLOR A color that is diametrically opposed in the range, or circle, of colors in the spectrum; for example, cyan (blue) is the complement of red.

COMPOSITE A material made from solid threads in a liquid matrix that is allowed to set.

COMPOUND A substance made from two or more elements that have chemically combined.

Ammonia is an example of a compound made from the elements hydrogen and nitrogen.

CONDENSE/CONDENSATION (1) To make something more concentrated or compact.

(2) The change of state from a gas or vapor to a liquid.

CONDUCTION The transfer of heat between two objects when they touch.

CONSTELLATION One of many commonly recognized patterns of stars in the sky.

CONVECTION/CONVECTION CURRENTS The circulating flow in a fluid (liquid or gas) that occurs when it is heated from below.

Convective flow is caused in a fluid by the tendency for hotter, and therefore less dense, material to rise and for colder, and therefore more dense, material, to sink with gravity. That results in a heat transfer.

CORE The central region of a body.

The core of the Earth is about 3,300 km in radius, compared with the radius of the whole Earth, which is 6,300 km.

CORONA (pl. **CORONAE**) (1) A colored circle seen around a bright object such as a star.

(2) The gases surrounding a star such as the Sun. In the case of the Sun and certain other stars these gases are extremely hot.

(3) A circular to oval pattern of faults, fractures, and ridges with a sagging center as found on Venus. In the case of Venus they are a few hundred kilometers in diameter.

CORONAL MASS EJECTIONS Very large bubbles of plasma escaping into the corona.

CORROSIVE SUBSTANCE Something that chemically eats away something else.

COSMOLOGICAL PRINCIPLE States that the way you see the universe is independent of the place where you are (your location). In effect, it means that the universe is roughly uniform throughout.

COSMONAUT A Russian space person.

COSMOS The universe and everything in it. The word "cosmos" suggests that the universe operates according to orderly principles.

CRATER A deep bowl-shaped depression in the surface of a body formed by the high-speed impact of another, smaller body.

Most craters are formed by the impact of asteroids and meteoroids. Craters have both a depression, or pit, and also an elevated rim formed of the material displaced from the central pit.

CRESCENT The appearance of the Moon when it is between a new Moon and a half Moon.

CRUST The solid outer surface of a rocky body.

The crust of the Earth is mainly just a few tens of kilometers thick, compared to the total radius of 6,300 km for the whole Earth. It forms much of the lithosphere.

CRYSTAL An ordered arrangement of molecules in a compound. Crystals that grow freely develop flat surfaces.

CYCLONE A large storm in which the atmosphere spirals inward and upward.

On Earth cyclones have a very low atmospheric pressure at their center and often contain deep clouds.

DARK MATTER Matter that does not shine or reflect light.

No one has ever found dark matter, but it is thought to exist because the amount of ordinary matter in the universe is not enough to account for many gravitational effects that have been observed.

DENSITY A measure of the amount of matter in a space.

Density is often measured in grams per cubic centimeter. The density of the Earth is 5.5 grams per cubic centimeter.

DEORBIT To move out of an orbital position and begin a reentry path toward the Earth.

DEPRESSION (1) A sunken area or hollow in a surface or landscape.

(2) A region of inward swirling air in the atmosphere associated with cloudy weather and rain.

DIFFRACTION The bending of light as it goes through materials of different density.

DISK A shape or surface that looks round and flat.

DOCK To meet with and attach to another space vehicle.

DOCKING PORT/STATION A place on the side of a spacecraft that contains some form of anchoring mechanism and an airlock.

DOPPLER EFFECT The apparent change in pitch of a fast-moving object as it approaches or leaves an observer.

DOWNLINK A communication to Earth from a spacecraft.

DRAG A force that hinders the movement of something.

DWARF STAR A star that shines with a brightness that is average or below.

EARTH The third planet from the Sun and the one on which we live.

The Earth belongs to the group of rocky planets. It is unique in having an oxygen-rich atmosphere and water, commonly found in its three phases—solid, liquid, and gas.

EARTHQUAKE The shock waves produced by the sudden movement of two pieces of brittle crust.

ECCENTRIC A noncircular, or oval, orbit.

ECLIPSE The time when light is cut off by a body coming between the observer and the source of the illumination (for example, eclipse of the Sun), or when the body the observer is on comes between the source of illumination and another body (for example, eclipse of the Moon).

It happens when three bodies are in a line. This phenomenon is not necessarily called an eclipse. Occultations of stars by the Moon and transits of Venus or Mercury are examples of different expressions used instead of "eclipse."

See also: **TOTAL ECLIPSE**.

ECOLOGY The study of living things in their environment.

ELECTRONS Negatively charged particles that are parts of atoms.

ELEMENT A substance that cannot be decomposed into simpler substances by chemical means.

Elements are the building blocks of compounds. For example, silicon and oxygen are elements. They combine to form the compound silicon dioxide, or quartz.

ELLIPTICAL GALAXY A galaxy that has an oval shape rather like a football, and that has no spiral arms.

EL NIÑO A time when ocean currents in the Pacific Ocean reverse from their normal pattern and disrupt global weather patterns. It occurs once every 4 or 5 years.

EMISSION Something that is sent or let out.

ENCKE GAP A gap between rings around Saturn named for the astronomer Johann Franz Encke (1791–1865).

EPOXY RESIN Adhesives that develop their strength as they react, or "cure," after mixing.

EQUATOR The ring drawn around a body midway between the poles.

EQUILIBRIUM A state of balance.

ESA The European Space Agency. ESA is an organizaton of European countries for cooperation in space research and technology. It operates several installations around Europe and has its headquarters in Paris, France.

ESCARPMENT A sharp-edged ridge.

EVAPORATE/EVAPORATION The change in state from liquid to a gas.

EXOSPHERE The outer part of the atmosphere starting about 500 km from the surface. This layer contains so little air that molecules rarely collide.

EXTRAVEHICULAR ACTIVITY Any task performed by people outside the protected environment of a space vehicle's pressurized compartments. Extravehicular activities (EVA) include repairing equipment in the Space Shuttle bay.

FALSE COLOR The colors used to make the appearance of some property more obvious.

They are part of the computer generation of an image.

FAULT A place in the crust where rocks have fractured, and then one side has moved relative to the other.

A fault is caused by excessive pressure on brittle rocks.

FLUORESCENT Emitting the visible light produced by a substance when it is struck by invisible waves, such as ultraviolet waves.

FRACTURE A break in brittle rock.

FREQUENCY The number of complete cycles of (for example, radio) waves received per second.

FRICTION The force that resists two bodies that are in contact.

For example, the effect of the ocean waters moving as tides slows the Earth's rotation.

FUSION The joining of atomic nuclei to form heavier nuclei.

This process results in the release of huge amounts of energy.

GALAXY A system of stars and interstellar matter within the universe.

Galaxies may contain billions of stars.

GALILEAN SATELLITES The four large satellites of Jupiter discovered by astronomer Galileo Galilei in 1610. They are Callisto, Europa, Ganymede, and Io.

GALILEO A U.S. space probe launched in October 1989 and designed for intensive investigation of Jupiter.

GEIGER TUBE A device to detect radioactive materials.

GEOSTATIONARY ORBIT A circular orbit 35,786 km directly above the Earth's equator.

Communications satellites frequently use this orbit. A satellite in a geostationary orbit will move at the same rate as the Earth's rotation, completing one revolution in 24 hours. That way it remains at the same point over the Earth's equator.

GEOSTATIONARY SATELLITE A man-made satellite in a fixed or geosynchronous orbit around the Earth.

GEOSYNCHRONOUS ORBIT An orbit in which a satellite makes one circuit of the Earth in 24 hours.

A geosynchronous orbit coincides with the Earth's orbit—it takes the same time to

complete an orbit as it does for the Earth to make one complete rotation. If the orbit is circular and above the equator, then the satellite remains over one particular point of the equator; that is called a geostationary orbit.

GEOSYNCLINE A large downward sag or trench that forms in the Earth's crust as a result of colliding tectonic plates.

GEYSER A periodic fountain of material. On Earth geysers are of water and steam, but on other planets and moons they are formed from other substances, for example, nitrogen gas on Triton.

GIBBOUS When between half and a full disk of a body can be seen lighted by the Sun.

GIMBALS A framework that allows anything inside it to move in a variety of directions.

GLOBAL POSITIONING SYSTEM A network of geostationary satellites that can be used to locate the position of any object on the Earth's surface.

GRANULATION The speckled pattern we see in the Sun's photosphere as a result of convectional overturning of gases.

GRAVITATIONAL FIELD The region surrounding a body in which that body's gravitational force can be felt.

The gravitational field of the Sun spreads over the entire solar system. The gravitational fields of the planets each exert some influence on the orbits of their neighbors.

GRAVITY/GRAVITATIONAL FORCE/ GRAVITATIONAL PULL The force of attraction between bodies. The larger an object, the more its gravitational pull on other objects.

The Sun's gravity is the most powerful in the solar system, keeping all of the planets and other materials within the solar system.

GREAT RED SPOT A large, almost permanent feature of the Jovian atmosphere that moves around the planet at about latitude 23°S.

GREENHOUSE EFFECT The increase in atmospheric temperature produced by the presence of carbon dioxide in the air.

Carbon dioxide has the ability to soak up heat radiated from the surface of a planet and partly prevent its escape. The effect is similar to that produced by a greenhouse.

GROUND STATION A receiving and transmitting station in direct communication with satellites. Such stations are characterized by having large dish-shaped antennae.

GULLY (pl. **GULLIES**) A trench in the land surface formed, on Earth, by running water.

GYROSCOPE A device in which a rapidly spinning wheel is held in a frame in such a way that it can rotate in any direction. The momentum of the wheel means that the gyroscope retains its position even when the frame is tilted.

HEAT SHIELD A protective device on the outside of a space vehicle that absorbs the heat during reentry and protects it from burning up.

HELIOPAUSE The edge of the heliosphere.

HELIOSEISMOLOGY The study of the internal structure of the Sun by modeling the Sun's patterns of internal shock waves.

HELIOSPHERE The entire range of influence of the Sun. It extends to the edge of the solar system.

HUBBLE SPACE TELESCOPE An orbiting telescope (and so a satellite) that was placed above the Earth's atmosphere so that it could take images that were far clearer than anything that could be obtained from the surface of the Earth.

HURRICANE A very violent cyclone that begins close to the equator, and that contains winds of over 117 km/hr.

ICE CAP A small mountainous region that is covered in ice.

INFRARED Radiation with a wavelength that is longer than red light.

INNER PLANETS The rocky planets closest to the Sun. They are Mercury, Venus, Earth, and Mars.

INTERNATIONAL SPACE STATION The international orbiting space laboratory.

INTERPLANETARY DUST The fine dustlike material that lies scattered through space, and that exists between the planets as well as in outer space.

INTERSTELLAR Between the stars.

IONIZED Matter that has been converted into small charged particles called ions.

An atom that has gained or lost an electron.

IONOSPHERE A part of the Earth's atmosphere in which the number of ions (electrically charged particles) is enough to affect how radio waves move.

The ionosphere begins about 50 km above the Earth's surface.

IRREGULAR SATELLITES Satellites that orbit in the opposite direction from their parent planet.

This motion is also called retrograde rotation.

ISOTOPE Atoms that have the same number of protons in their nucleus, but that have different masses; for example, carbon-12 and carbon-14.

JOVIAN PLANETS An alternative group name for the gas giant planets: Jupiter, Saturn, Uranus, and Neptune.

JUPITER The fifth planet from the Sun and two planets farther away from the Sun than the Earth.

Jupiter is 318 times as massive as the Earth and 1,500 times as big by volume. It is the largest of the gas giants.

K Named for British scientist Lord Kelvin (1824–1907), it is a measurement of absolute temperature. Zero K is called absolute zero and is only approached in deep space: ice melts at 273 K, and water boils at 373 K.

KEELER GAP A gap in the rings of Saturn named for the astronomer James Edward Keeler (1857–1900).

KILOPARSEC A unit of a thousand parsecs. A parsec is the unit used for measuring the largest distances in the universe.

KUIPER BELT A belt of planetesimals (small rocky bodies, one kilometer to hundreds of kilometers across) much closer to the Sun than the Oort cloud.

LANDSLIDE A sudden collapse of material on a steep slope.

LA NIÑA Below normal ocean temperatures in the eastern Pacific Ocean that disrupt global weather patterns.

LATITUDE Angular distance north or south of the equator, measured through 90°.

LAUNCH VEHICLE/LAUNCHER A system of propellant tanks and rocket motors or engines designed to lift a payload into space. It may, or may not, be part of a space vehicle.

LAVA Hot, melted rock from a volcano.

Lava flows onto the surface of a planet and cools and hardens to form new rock. Most of the lava on Earth is made of basalt.

LAVA FLOW A river or sheet of liquid volcanic rock.

LAWS OF MOTION Formulated by Sir Isaac Newton, they describe the forces that act on a moving object.

The first law states that an object will keep moving in a straight line at constant speed unless it is acted on by a force.

The second law states that the force on an object is related to the mass of the object multiplied by its acceleration.

The third law states that an action always has an equal and directly opposite reaction.

LIFT An upthrust on the wing of a plane that occurs when it moves rapidly through the air. It is the main way of suspending an airplane during flight. The engines simply provide the forward thrust.

LIGHT-YEAR The distance traveled by light through space in one Earth year, or 63,240 astronomical units.

The speed of light is the speed that light travels through a vacuum, which is 299,792 km/s.

LIMB The outer edge of a celestial body, including an atmosphere if it has one.

LITHOSPHERE The upper part of the Earth, corresponding generally to the crust and believed to be about 80 km thick.

LOCAL GROUP The Milky Way, the Magellanic Clouds, the Andromeda Galaxy, and over 20 other relatively near galaxies.

LUNAR Anything to do with the Moon.

MAGELLANIC CLOUD Either of two small galaxies that are companions to the Milky Way Galaxy.

MAGMA Hot, melted rock inside the Earth that, when cooled, forms igneous rock.

Magma is associated with volcanic activity.

MAGNETIC FIELD The region of influence of a magnetic body.

The Earth's magnetic field stretches out beyond the atmosphere into space. There it interacts with the solar wind to produce auroras.

MAGNETISM An invisible force that has the property of attracting iron and similar metals.

MAGNETOPAUSE The outer edge of the magnetosphere.

MAGNETOSPHERE A region in the upper atmosphere, or around a planet, where magnetic phenomena such as auroras are found.

MAGNITUDE A measure of the brightness of a star.

The apparent magnitude is the brightness of a celestial object as seen from the Earth. The absolute magnitude is the standardized brightness measured as though all objects were the same distance from the Earth. The brighter the object, the lower its magnitude number. For example, a star of magnitude 4 is 2.5 times as bright as one of magnitude 5. A difference of five magnitudes is the same as a difference in brightness of 100 to 1. The brightest stars have negative numbers. The Sun's apparent magnitude is −26.8. Its absolute magnitude is 4.8.

MAIN SEQUENCE The 90% of stars in the universe that represent the mature phase of stars with small or medium mass.

MANTLE The region of a planet between the core and the crust.

The Earth's mantle is about 2,900 km thick, and its upper surface may be molten in some places.

MARE (pl. MARIA) A flat, dark plain created by lava flows. They were once thought to be seas.

MARS The fourth planet from the Sun in our solar system and one planet farther away from the Sun than the Earth.

Mars is a rocky planet almost half the diameter of Earth that is a distinctive rust-red color.

MASCON A region of higher surface density on the Moon.

MASS The amount of matter in an object.

The amount of matter, and so the mass, remains the same, but the effect of gravity gives the mass a weight. The weight depends on the gravitational pull. Thus a ball will have the same mass on the Earth and on the Moon, but it will weigh a sixth as much on the Moon because the force of gravity there is only a sixth as strong.

MATTER Anything that exists in physical form.

Everything we can see is made of matter. The building blocks of matter are atoms.

MERCURY The closest planet to the Sun in our solar system and two planets closer to the Sun than Earth.

Mercury is a gray-colored rocky planet less than half the diameter of Earth. It has the most extreme temperature range of any planet in our solar system.

MESOSPHERE One of the upper regions of the atmosphere, beginning at the top of the stratosphere and continuing from 50 km upward until the temperature stops declining.

METEOR A streak of light (shooting star) produced by a meteoroid as it enters the Earth's atmosphere.

The friction with the Earth's atmosphere causes the small body to glow (become incandescent). That is what we see as a streak of light.

METEORITE A meteor that reaches the Earth's surface.

METEOROID A small body moving in the solar system that becomes a meteor if it enters the Earth's atmosphere.

Meteoroids are typically only a few millimeters across and burn up as they go through the atmosphere, but some have crashed to the Earth, making large craters.

MICROMETEORITES Tiny pieces of space dust moving at high speeds.

MICRON A millionth of a meter.

MICROWAVELENGTH Waves at the shortest end of the radio wavelengths.

MICROWAVE RADIATION The background radiation that is found everywhere in space, and whose existence is used to support the Big Bang theory.

MILKY WAY The spiral galaxy in which our star and solar system are situated.

MINERAL A solid crystalline substance.

MINOR PLANET Another term for an asteroid.

M NUMBER In 1781 Charles Messier began a catalogue of the objects he could see in the night sky. He gave each of them a unique number. The first entry was called M1. There is no significance to the number in terms of brightness, size, closeness, or otherwise.

MODULE A section, or part, of a space vehicle.

MOLECULE A group of two or more atoms held together by chemical bonds.

MOLTEN Liquid, suggesting that it has changed from a solid.

MOMENTUM The mass of an object multiplied by its velocity.

MOON The natural satellite that orbits the Earth.

Other planets have large satellites, or moons, but none is relatively as large as our Moon, suggesting that it has a unique origin.

MOON The name generally given to any large natural satellite of a planet.

MOUNTAIN RANGE A long, narrow region of very high land that contains several or many mountains.

NASA The National Aeronautics and Space Administration.

NASA was founded in 1958 for aeronautical and space exploration. It operates several installations around the country and has its headquarters in Washington, D.C.

NEAP TIDE A tide showing the smallest difference between high and low tides.

NEBULA (pl. NEBULAE) Clouds of gas and dust that exist in the space between stars.

The word means mist or cloud and is also used as an alternative to galaxy. The gas makes up to 5% of the mass of a galaxy. What a nebula looks like depends on the arrangement of gas and dust within it.

NEPTUNE The eighth planet from the Sun in our solar system and five planets farther away from the Sun than the Earth.

Neptune is a gas planet that is almost four times the diameter of Earth. It is blue.

NEUTRINOS An uncharged fundamental particle that is thought to have no mass.

NEUTRONS Particles inside the core of an atom that are neutral (have no charge).

NEUTRON STAR A very dense star that consists only of tightly packed neutrons. It is the result of the collapse of a massive star.

NOBLE GASES The unreactive gases, such as neon, xenon, and krypton.

NOVA (pl. NOVAE) (1) A star that suddenly becomes much brighter, then fades away to its original brightness within a few months.
See also: **SUPERNOVA**.

(2) A radiating pattern of faults and fractures unique to Venus.

NUCLEAR DEVICES Anything that is powered by a source of radioactivity.

NUCLEUS (pl. NUCLEI) The centermost part of something, the core.

OORT CLOUD A region on the edge of the solar system that consists of planetesimals and comets that did not get caught up in planet making.

OPTICAL Relating to the use of light.

ORBIT The path followed by one object as it tracks around another.

The orbits of the planets around the Sun and moons around their planets are oval, or elliptical.

ORGANIC MATERIAL Any matter that contains carbon and is alive.

OUTER PLANETS The gas giant planets Jupiter, Saturn, Uranus, and Neptune plus the rocky planet Pluto.

OXIDIZER The substance in a reaction that removes electrons from and thereby oxidizes (burns) another substance.

In the case of oxygen this results in the other substance combining with the oxygen to form an oxide (also called an oxidizing agent).

OZONE A form of oxygen (O_3) with three atoms in each molecule instead of the more usual two (O_2).

OZONE HOLE The observed lack of the gas ozone in the upper atmosphere.

PARSEC The unit used for measuring the largest distances in the universe.

A parsec is the distance at which an observer in space would see the radius of the orbit as making one second of arc. This gives a distance of about 3.26 light-years.
See also: **KILOPARSEC**.

PAYLOAD The spacecraft that is carried into space by a launcher.

PENUMBRA (1) A region that is in semidarkness during an eclipse.

(2) The part of a sunspot surrounding the umbra.

PERCOLATE To flow by gravity between particles, for example, of soil.

PERIGEE The point on an orbit where the orbiting object is as close as it ever comes to the object it is orbiting.

PHARMACEUTICAL Relating to medicinal drugs.

PHASE The differing appearance of a body that is closer to the Sun, and that is illuminated by it.

PHOTOCHEMICAL SMOG A hazy atmosphere, often brown, resulting from the reaction of nitrogen gases with sunlight.

PHOTOMOSAIC A composite picture made up of several other pictures that individually only cover a small area.

PHOTON A particle (quantum) of electromagnetic radiation.

PHOTOSPHERE A shell of the Sun that we regard as its visible surface.

PHOTOSYNTHESIS The process that plants use to combine the substances in the environment, such as carbon dioxide, minerals, and water, with oxygen and energy-rich organic compounds by using the energy of sunlight.

PIONEER A name for a series of unmanned U.S. spacecraft.

Pioneer 1 was launched into lunar orbit on October 11, 1958. The others all went into deep space.

PLAIN A flat or gently rolling part of a landscape.

Plains are confined to lowlands. If a flat surface exists in an upland, it is called a plateau.

PLANE A flat surface.

PLANET Any of the large bodies that orbit the Sun.

The planets are (outward from the Sun): Mercury, Venus, Earth, Mars, Jupiter, Saturn, Uranus, Neptune, and Pluto. The rocky planets all have densities greater than 3 grams per cubic centimeter; the gaseous ones less than 2 grams per cubic centimeter.

PLANETARY NEBULA A compact ring or oval nebula that is made of material thrown out of a hot star.

The term "planetary nebula" is a misnomer; dying stars create these cocoons when they lose outer layers of gas. The process has nothing to do with planet formation, which is predicted to happen early in a star's life.

The term originates from a time when people, looking through weak telescopes, thought that the nebulae resembled planets within the solar system, when in fact they were expanding shells of glowing gas in far-off galaxies.

PLANETESIMAL Small rocky bodies one kilometer to hundreds of kilometers across.

The word especially relates to materials that exist in the early stages of the formation of a star and its planets from the dust of a nebula, which will eventually group together to form planets. Some are rock, others a mixture of rock and ice.

PLANKTON Microscopic creatures that float in water.

PLASMA A collection of charged particles that behaves something like a gas. It can conduct an electric charge and be affected by magnetic fields.

PLASTIC The ability of certain solid substances to be molded or deformed to a new shape under pressure without cracking.

PLATE A very large unbroken part of the crust of a planet. Also called tectonic plate.

On Earth the tectonic plates are dragged across the surface by convection currents in the underlying mantle.

PLATEAU An upland plain or tableland.

PLUTO The ninth planet from the Sun and six planets farther from the Sun than the Earth.

Pluto is one of the rocky planets, but it is very different from the others, perhaps being a mixture of rock and ice. It is about two-thirds the size of our Moon.

POLE The geographic pole is the place where a line drawn along the axis of rotation exits from a body's surface.

Magnetic poles do not always correspond with geographic poles.

POLYMER A compound that is made up of long chains formed by combining molecules called monomers as repeating units. ("Poly" means many, "mer" means part.)

PRESSURE The force per unit area.

PROBE An unmanned spacecraft designed to explore our solar system and beyond.

Voyager, Cassini, and Magellan are examples of probes.

PROJECTILE An object propelled through the air or space by an external force or an on-board engine.

PROMINENCE A cloud of burning ionized gas that rises through the Sun's chromosphere into the corona. It can take the form of a sheet or a loop.

PROPELLANT A gas, liquid, or solid that can be expelled rapidly from the end of an object in order to give it motion.

Liquefied gases and solids are used as rocket propellants.

PROPULSION SYSTEM The motors or rockets and their tanks designed to give a launcher or space vehicle the thrust it needs.

PROTEIN Molecules in living things that are vital for building tissues.

PROTONS Positively charged particles from the core of an atom.

PROTOSTAR A cloud of gas and dust that begins to swirl around; the resulting gravity gives birth to a star.

PULSAR A neutron star that is spinning around, releasing electromagnetic radiation, including radio waves.

QUANTUM THEORY A concept of how energy can be divided into tiny pieces called quanta, which is the key to how the smallest particles work and how they build together to make the universe around us.

QUASAR A rare starlike object of enormous brightness that gives out radio waves, which are thought to be released as material is sucked toward a black hole.

RADAR Short for radio detecting and ranging. A system of bouncing radio waves from objects in order to map their surfaces and find out how far away they are.

Radar is useful in conditions where visible light cannot be used.

RADIATION/RADIATE The transfer of energy in the form of waves (such as light and heat) or particles (such as from radioactive decay of a material).

RADIOACTIVE/RADIOACTIVITY The property of some materials that emit radiation or energetic particles from the nucleus of their atoms.

RADIOACTIVE DECAY The change that takes place inside radioactive materials and causes them to give out progressively less radiation over time.

RADIO GALAXY A galaxy that gives out radio waves of enormous power.

RADIO INTERFERENCE Reduction in the radio communication effectiveness of the ionosphere caused by sunspots and other increases in the solar wind.

RADIO TELESCOPE A telescope that is designed to detect radio waves rather than light waves.

RADIO WAVES A form of electromagnetic radiation, like light and heat. Radio waves have a longer wavelength than light waves.

RADIUS (pl. **RADII**) The distance from the center to the outside of a circle or sphere.

RAY A line across the surface of a planet or moon made by material from a crater being flung across the surface.

REACTION An opposition to a force.

REACTIVE The ability of a chemical substance to combine readily with other substances. Oxygen is an example of a reactive substance.

RED GIANT A cool, large, bright star at least 25 times the diameter of our Sun.

REFLECT/REFLECTION/REFLECTIVE To bounce back any light that falls on a surface.

REGULAR SATELLITES Satellites that orbit in the same direction as their parent planet. This motion is also called synchronous rotation.

RESOLVING POWER The ability of an optical telescope to form an image of a distant object.

RETROGRADE DIRECTION An orbit the opposite of normal—that is, a planet that spins so the Sun rises in the west and sinks in the east.

RETROROCKET A rocket that fires against the direction of travel in order to slow down a space vehicle.

RIDGE A narrow crest of an upland area.

RIFT A trench made by the sinking of a part of the crust between parallel faults.

RIFT VALLEY A long trench in the surface of a planet produced by the collapse of the crust in a narrow zone.

ROCKET Any kind of device that uses the principle of jet propulsion, that is, the rapid release of gases designed to propel an object rapidly.

The word is also applied loosely to fireworks and spacecraft launch vehicles.

ROCKET ENGINE A propulsion system that burns liquid fuel such as liquid hydrogen.

ROCKET MOTOR A propulsion system that burns solid fuel such as hydrazine.

ROCKETRY Experimentation with rockets.

ROTATION Spinning around an axis.

SAND DUNE An aerodynamically shaped hump of sand.

SAROS CYCLE The interval of 18 years 11$^1/_3$ days needed for the Earth, Sun, and Moon to come back into the same relative positions. It controls the pattern of eclipses.

SATELLITE (1) An object that is in an orbit around another object, usually a planet.

The Moon is a satellite of the Earth.
See also: **IRREGULAR SATELLITE, MOON, GALILEAN SATELLITE, REGULAR SATELLITE, SHEPHERD SATELLITE.**

(2) A man-made object that orbits the Earth. Usually used as a term for an unmanned spacecraft whose job is to acquire or transfer data to and from the ground.

SATURN The sixth planet from the Sun and three planets farther away from the Sun than the Earth.

It is the least-dense planet in the solar system, having 95 times the mass of the Earth, but 766 times the volume. It is one of the gas giant planets.

SCARP The steep slope of a sharp-crested ridge.

SEASONS The characteristic cycle of events in the heating of the Earth that causes related changes in weather patterns.

SEDIMENT Any particles of material that settle out, usually in layers, from a moving fluid such as air or water.

SEDIMENTARY Rocks deposited in layers.

SEISMIC Shaking, relating to earthquakes.

SENSOR A device used to detect something. Your eyes, ears, and nose are all sensors. Satellites use sensors that mainly detect changes in radio and other waves, including sunlight.

SHEPHERD SATELLITES Larger natural satellites that have an influence on small debris in nearby rings because of their gravity.

SHIELD VOLCANO A volcanic cone that is broad and gently sloping.

SIDEREAL MONTH The average time that the Moon takes to return to the same position against the background of stars.

SILT Particles with a range of 2 microns to 60 microns across.

SLINGSHOT TRAJECTORY A path chosen to use the attractive force of gravity to increase the speed of a spacecraft.

The craft is flown toward the planet or star, and it speeds up under the gravitational force. At the correct moment the path is taken to send the spacecraft into orbit and, when pointing in the right direction, to turn it from orbit, with its increased velocity, toward the final destination.

SOLAR Anything to do with the Sun.

SOLAR CELL A photoelectric device that converts the energy from the Sun (solar radiation) into electrical energy.

SOLAR FLARE Any sudden explosion from the surface of the Sun that sends ultraviolet radiation into the chromosphere. It also sends out some particles that reach Earth and disrupt radio communications.

SOLAR PANELS Large flat surfaces covered with thousands of small photoelectric devices that convert solar radiation into electricity.

SOLAR RADIATION The light and heat energy sent into space from the Sun.

Visible light and heat are just two of the many forms of energy sent by the Sun to the Earth.

SOLAR SYSTEM The Sun and the bodies orbiting around it.

The solar system contains nine major planets, at least 60 moons (large natural satellites), and a vast number of asteroids and comets, together with the gases within the system.

SOLAR WIND The flow of tiny charged particles (called plasma) outward from the Sun.

The solar wind stretches out across the solar system.

SONIC BOOM The noise created when an object moves faster than the speed of sound.

SPACE Everything beyond the Earth's atmosphere.

The word "space" is used rather generally. It can be divided up into inner space—the solar system, and outer space—everything beyond the solar system, for example, interstellar space.

SPACECRAFT Anything capable of moving beyond the Earth's atmosphere. Spacecraft can be manned or unmanned. Unmanned spacecraft are often referred to as space probes if they are exploring new areas.

SPACE RACE The period from the 1950s to the 1970s when the United States and the Soviet Union competed to be first in achievements in space.

SPACE SHUTTLE NASA's reusable space vehicle that is launched like a rocket but returns like a glider.

SPACE STATION A large man-made satellite used as a base for operations in space.

SPEED OF LIGHT *See:* **LIGHT-YEAR.**

SPHERE A ball-shaped object.

SPICULES Jets of relatively cool gas that move upward through the chromosphere into the corona.

SPIRAL GALAXY A galaxy that has a core of stars at the center of long curved arms made of even more stars arranged in a spiral shape.

SPRING TIDE A tide showing the greatest difference between high and low tides.

STAR A large ball of gases that radiates light. The star nearest the Earth is the Sun.

There are enormous numbers of stars in the universe, but few can be seen with the naked eye. Stars may occur singly, as our Sun, or in groups, of which pairs are most common.

STAR CLUSTER A group of gravitationally connected stars.

STELLAR WIND The flow of tiny charged particles (called plasma) outward from a star.

In our solar system the stellar wind is the same as the solar wind.

STRATOSPHERE The region immediately above the troposphere where the temperature increases with height, and the air is always stable.

It acts like an invisible lid, keeping the clouds in the troposphere.

SUBDUCTION ZONES Long, relatively thin, but very deep regions of the crust where one plate moves down and under, or subducts, another. They are the source of mountain ranges.

SUN The star that the planets of the solar system revolve around.

The Sun is 150 million km from the Earth and provides energy (in the form of light and heat) to our planet. Its density of 1.4 grams per cubic centimeter is similar to that of a gas giant planet.

SUNSPOT A spiral of gas found on the Sun that is moving slowly upward, and that is cooler than the surrounding gas and so looks darker.

SUPERNOVA A violently exploding star that becomes millions or even billions of times brighter than when it was younger and stable.
See also: **NOVA.**

SYNCHRONOUS Taking place at the same time.

SYNCHRONOUS ORBIT An orbit in which a satellite (such as a moon) moves around a planet in the same time that it takes for the planet to make one rotation on its axis.

SYNCHRONOUS ROTATION When two bodies make a complete rotation on their axes in the same time.

As a result, each body always has the same side facing the other. The Moon and Venus are in synchronous rotation with the Earth.

SYNODIC MONTH The complete cycle of phases of the Moon as seen from Earth. It is 29.531 solar days (29 days, 12 hours, 44 minutes, 3 seconds).

SYNODIC PERIOD The time needed for an object within the solar system, such as a planet, to return to the same place relative to the Sun as seen from the Earth.

TANGENT A direction at right angles to a line radiating from a circle or sphere.

If you make a wheel spin, for example, by repeatedly giving it a glancing blow with your hand, the glancing blow is moving along a tangent.

TELECOMMUNICATIONS Sending messages by means of telemetry, using signals made into waves such as radio waves.

THEORY OF RELATIVITY A theory based on how physical laws change when an observer is moving. Its most famous equation says that at the speed of light, energy is related to mass and the speed of light.

THERMOSPHERE A region of the upper atmosphere above the mesosphere.

It absorbs ultraviolet radiation and is where the ionosphere has most effect.

THRUST A very strong and continued pressure.

THRUSTER A term for a small rocket engine.

TIDE Any kind of regular, or cyclic, change that occurs due to the effect of the gravity of one body on another.

We are used to the ocean waters of the Earth being affected by the gravitational pull of the Moon, but tides also cause a small alteration of the shape of a body. This is important in determining the shape of many moons and may even be a source of heating in some.
See also: **NEAP TIDE** and **SPRING TIDE.**

TOPOGRAPHY The shape of the land surface in terms of height.

TOTAL ECLIPSE When one body (such as the Moon or Earth) completely obscures the light source from another body (such as the Earth or Moon).

A total eclipse of the Sun occurs when it is completely blocked out by the Moon.

A total eclipse of the Moon occurs when it passes into the Earth's shadow to such a degree that light from the Sun is completely blocked out.

TRAJECTORY The curved path followed by a projectile.
See also: **SLINGSHOT TRAJECTORY.**

TRANSPONDER Wireless receiver and transmitter.

TROPOSPHERE The lowest region of the atmosphere, where all of the Earth's clouds form.

TRUSS Tubing arrayed in the form of triangles and designed to make a strong frame.

ULTRAVIOLET A form of radiation that is just beyond the violet end of the visible spectrum and so is called "ultra" (more than) violet. At the other end of the visible spectrum is "infra" (less than) red.

UMBRA (1) A region that is in complete darkness during an eclipse.
(2) The darkest region in the center of a sunspot.

UNIVERSE The entirety of everything there is; the cosmos.

Many space scientists prefer to use the term "cosmos," referring to the entirety of energy and matter.

UNSTABLE In atmospheric terms the potential churning of the air in the atmosphere as a result of air being heated from below. There is a chance of the warmed, less-dense air rising through the overlying colder, more-dense air.

UPLINK A communication from Earth to a spacecraft.

URANUS The seventh planet from the Sun and four planets farther from the Sun than the Earth.

Its diameter is four times that of the Earth. It is one of the gas giant planets.

VACUUM A space that is entirely empty.
A vacuum lacks any matter.

VALLEY A natural long depression in the landscape.

VELOCITY A more precise word to describe how something is moving, because movement has both a magnitude (speed) and a direction.

VENT The tube or fissure that allows volcanic materials to reach the surface of a planet.

VENUS The second planet from the Sun and our closest neighbor.

It appears as an evening and morning "star" in the sky. Venus is very similar to the Earth in size and mass.

VOLCANO A mound or mountain that is formed from ash or lava.

VOYAGER A pair of U.S. space probes designed to provide detailed information about the outer regions of the solar system.

Voyager 1 was launched on September 5, 1977. Voyager 2 was launched on August 20, 1977, but traveled more slowly than Voyager 1. Both Voyagers are expected to remain operational until 2020, by which time they will be well outside the solar system.

WATER CYCLE The continuous cycling of water, as vapor, liquid, and solid, between the oceans, the atmosphere, and the land.

WATER VAPOR The gaseous form of water. Also sometimes referred to as moisture.

WEATHERING The breaking down of a rock, perhaps by water, ice, or repeated heating and cooling.

WHITE DWARF Any star originally of low mass that has reached the end of its life.

X-RAY An invisible form of radiation that has extremely short wavelengths just beyond the ultraviolet.

X-rays can go through many materials that light will not.

SET INDEX

Using the set index

This index covers all eight volumes in the *Space Science* set:

Vol. no. Title
1: *How the universe works*
2: *Sun and solar system*
3: *Earth and Moon*
4: *Rocky planets*
5: *Gas giants*
6: *Journey into space*
7: *Shuttle to Space Station*
8: *What satellites see*

An example entry:
Index entries are listed alphabetically.

—————— /

Moon rover **3:** 48–49, **6:** 51

/ ——————

Volume numbers are in bold and are followed by page references.
In the example above, "Moon rover" appears in Volume 3: *Earth and Moon* on pages 48–49 and in Volume 6: *Journey into space* on page 51. Many terms are also covered separately in the Glossary on pages 58–64.

See, see also, or *see under* refers to another entry where there will be additional relevant information.